T0139303

Economics of Environment and Development

Economics of Environment and Development

Edited by

Pushpam Kumar

Taylor & Francis
Taylor & Francis Group
Boca Raton London New York

CRC is an imprint of the Taylor & Francis Group,
an informa business

Ane Books India

Economics of Environment and Development

© **Ane Books India**

First Published in 2008 by

Ane Books India

4821 Parwana Bhawan, 1st Floor
24 Ansari Road, Darya Ganj, New Delhi -110 002, India
Tel: +91 (011) 2327 6843-44, 2324 6385
Fax: +91 (011) 2327 6863
e-mail: anebooks@vsnl.com
Website: www.anebooks.com

For

CRC Press
Taylor & Francis Group
6000 Broken Sound Parkway, NW, Suite 300
Boca Raton, FL 33487 U.S.A.
Tel : 561 998 2541
Fax : 561 997 7249 or 561 998 2559
Web : www.taylorandfrancis.com

For distribution in rest of the world other than the Indian sub-continent

ISBN-10 : 1 42007 067 3
ISBN-13 : 978 1 42007 067 5

British Library Cataloguing in Publication Data
A catalogue record for this book is available from the British Library

Printed at Brijbasi Art Press, Noida

Contents

List of Contributors

Anderson, Kym
International Trade Unit of the World Bank's
Development Research Group,
Washington DC, USA

Bartelmus, Peter; Carsten Stahmer and Jan van Tongeren
United Nations Statstical Office; Federal Statistical Office,
Germany
and United Nations Statistical Office

Duraiappah, Anantha K.
International Institute of Sustainable Development,
Winnipeg, Canada

Hamilton, Kirk
Environment Department,
World bank,
Washington DC, USA

Krutilla, K.
School of Public and Environmental Affairs,
Indiana University, Bloomington,
USA

Kumar, Pushpam
Institute of Economic Growth
University of Delhi Enclave,
Delhi

Munasinghe, Mohan
Munasingle Institute for Development Colombo, Srilanka
and Environment Department, World bank, Washington DC,
USA

Murty, M.N.
Institute of Economic Growth
University of Delhi Enclave,
Delhi

Rauscher, M.
Faculty of Economics,
Universtät Rostock,
Germany

Söderbaum, Peter
Mä Lardalen University, Sweden

Spash, Clive L.
The Macaulay Institute, Craigiebucker Aberdeen, UK
and Geography and Environment Department University of
Aberdeen, UK

Steininger, Karl W.
Department of Economics,
University of Gray,
Austria

Foreword

The three pillars of sustainable development were widely agreed at the 1992 Earth Summit to be economic viability, environmental sustainability, and social equity. But the economic leg of the stool has been far longer and stronger than the other two legs, making for a rather uncomfortable perch. The emerging discipline of environmental and natural resource economics helps to design a better stool, where the three legs are in a more appropriate balance. This book, assembled by Pushpam Kumar, provides an excellent introduction into the key perspectives promoted by this critically important discipline.

Many new perspectives on links between environment and development are becoming available, and are discussed in this book. One of the most interesting is the concept of ecosystem services. Over the past several decades, the provisioning services—namely the production of food, logs, fish, and so forth—have received far greater attention than the ecosystem services that support processes such as watershed protection, conservation of pollinators, soil formation, carbon sequestration, and conservation of biodiversity. But now, new initiatives are seeking to capture some of these ecosystem services that previously remained outside the market place.

It is important to remember that we are dealing with dynamic systems, essentially a moving target. Climate change, growing economies, demographic changes, new technologies, and changes in global and national security are just a few of the changes that make the process of sustainable development such a challenge. But this book shows that dealing with environmental change in an economically sensible way offers numerous entry points into sustainable forms of development, that help people adapt to changing conditions.

Globalisation means more open borders for freer flows of finance, business, trade, ideas and cultural values. Clearly, embracing globalisation entails political, social, and cultural risks, requiring changes that may not always be welcomed by those in power. For example, the global economy requires transparency, accountability, and the rule of law, so it should not be surprising that some leaders, especially in Asia, have denounced globalisation as a new form of western imperialism, even as they reap huge financial gains from the process. While some people in South Korea, Thailand, Singapore and the Philippines have welcomed the opportunity for reform, other countries in Asia have been increasingly critical of the process. They perceive that the possibility of internal fragmentation as an externality of globalization is a real threat, especially in multiethnic states such as India, Indonesia and Malaysia.

The book also deals with so-called "externalities", the costs of an activity that are not paid by those undertaking the activity. Pollution is one obvious externality, where industrialists, miners, or farmers pass the costs of their pollution onto the public at large. One of the most interesting ways of dealing with pollution is through "internalising" the costs, in other words, making the polluter pay for his pollution. Of course, ultimately all costs are paid by the consumer, but by incorporating the cost of pollution into the prices paid by the consumer, a positive feedback loop between consumption and production can be established. This hopefully will drive new enterprises to be less polluting and hence more cost-effective.

Another way of externalising costs is seen in the field of international trade, where the drive to keep prices low is encouraging governments to externalise as many of the costs as possible. A global trading system makes it very difficult for the end consumers to have much of an idea about where the environmental impacts are actually being felt, the rural communities who live among the greatest biodiversity have great difficulty conserving their resources against the more powerful outsiders who have other plans for their lands and resources. A Parisian sipping her espresso in a sidewalk café has no idea about the environmental impact of growing coffee in the highlands of Guatemala or Sumatra. More effective feedback mechanisms will include labelling of sustainably-produced coffee and prices that better reflect the full environmental costs of production.

The book also considers the role of the commercial sector in the environment. One of the most promising developments of the early 21st century is the increasing interest of the private sector in promoting sustainable forms of development. Socially responsible investments rose by more than a third from 1999 to 2001, and have exceeded the US$2 trillion mark. Businesses have many reasons to embrace more effective management of biodiversity, ranging from more sustainable supplies of the natural resources upon which their businesses depend to enhanced reputation as a company that behaves in a socially responsible manner.

Finally, the book addresses the difficult topic of poverty. According to UNDP, per capita income is lower today than it was 20 years ago in over 70 countries. And between 1960 and 1995, the income of the 20% of the world's population living in the richest countries increased from 30 times greater than that of the 20% in the poorest countries to 82 times greater. A similar gap is also seen domestically in many countries. The gap between rich and poor is widening at an accelerating pace, which eventually will lead to growing social tension. Environmental economics can help address this tension.

I have just touched on a few of the many points that this excellent book covers in considerable detail. It provides an essential entry point for students of economics, development, environment, and other related fields, and hopefully will contribute to a more sustainable and prosperous India.

Jeffrey A. McNeely
Chief Scientist
IUCN-The World Conservation Union
Gland, Switzerland

Acknowledgements

The editor along with the publisher wishes to thank the authors and the following publishers who have given permission for the use of copyright material.

(1) Blackwell Publishers, for the paper titled, "Integrated Environmental and Economic Accounting: Framework for SNA Satellite System', by Peter Bartelmus; Carsten Stahmer and Jan Van Tongeren in *Review of Income and Wealth, 37 (2), June, 111-48.*

(2) Inderscience Enterprises Ltd. For the paper The sustainnomics trans-disciplinary meta-framework for making development more sustainable by Monan Munasinghe in *Int. Journal of sustainable Development, Vol 5, Nos. 1/2, 2002*

(3) Elsevier Science Ltd. For the paper Poverty and environmental degradation: A review and analysis of the nexus by A K Duraiappah in *World Development* Vol 26, No.12, 1998.

(4) Edward Elgar for the paper-Formal models and practical measurement for greening the accounts by Kirk Hamilton in Sandrine Simon and John Proops edited (1998) Greening the Accounts.

(5) Harvester Wheatsheaf for paper-The Standard Welfare Economics of Policies Affecting Trade and the Environment by Kym Anderson in Kym Anderson and Richard Blackhurst edited (1994) The Greening of World Trade Issues.

(6) *Handbook of Environmental and Resource Economics,* edited by Jeroen C.J.M van den Bergh, Edward Elgar, for the following articles.

   (a) Environmental Policy in Open Economics, by Michael Rauscher
   (b) Partial Equilibrium Models of Trade and the Environment, by Kerry Krutila
   (c) General Models of Environmental Policy and Foreign Trade, by Karl W. Steininger

I would like to express my sincere thanks to Dr Jeffrey McNeeley, Chief Scientist, IUCN for taking time to read the assembled articles and write the foreword. I am also thankful to Mr. Mridu Prabal Goswami; a Senior Research Analyst with the Institute of Economic Growth who ably supported me in completing the work especially the chapter on Environmental Management in Business Firm. Ashish Bharadwaj cheerfully volunteered to help me in various ways which really facilitated the work, he deserves a special thank.

Finally, my thanks are due to Mr Sunil Saxena of the Ane Books, who took special interest in this project and this book could become the reality.

Every effort has been made to trace all the copyright holders but if any has been inadvertently overlooked, the publishers will be pleased to make the necessary arrangements for change at the first opportunity.

Pushpam Kumar

1

Introduction

Pushpam Kumar

The discipline of environmental and natural resource economics has evolved rapidly in recent years. It has emerged not only as an academic subject for teaching and research in the universities and research institutions, but also as a tool in the hands of decision makers and planners to solve the problems of environmental depletion and degradation. Application of various economic theories and design of instruments for environmental governance has yielded (and promises to do so in the future too) solutions to problems like air and water pollution, loss of bio-diversity and perturbances in to the ecosystems health, to name a few.

In the past, economists like David Ricardo, Malthus and Mills in 18th and 19th century have discussed the issues of land degradation, population growth and utility of pristine wilderness respectively. But their concerns could only get crystallized in the early 20th century through the works of Pigou (1920), Hotelling 1939, Clark (1976) and Boulding(1978). By this time, the foundation of neoclassical economics had been laid by Menger and Marshall and it had got further strengthened too. And so, issues of pollution, management of natural resources like forestry, fisheries and mining etc. were addressed by various economists in the paradigm of neoclassical economic theory. It was understood that economic activity uses natural resources and also returns waste to nature and more importantly, the scale of that activity is determined independently of the rate of replacement of resources or the waste absorption capacity of the environment. At low levels of economic activity, this asymmetry goes unnoticed because the inputs required to be extracted are small. Traditional microeconomics treats environmental effects as externalities of production or consumption (Pigou 1920 and Ayres and Kneese 1969). As long as the uncompensated technological externalities of the production and consumption processes are small, the environment, principally water and air absorb the residuals of these activities.

Beyond certain scales of economic activity, the pervasive nature of these externalities and limits on the waste assimilative capacity of environmental sinks results in the need for such residues to be treated as a part of the materials balancing problem in the economy. The problem of externality management thus moves from the realm of microeconomics to that of macroeconomic management of the flow of materials in an input-output framework. The presence of waste generating externalities imposes a constraint on production levels in such a framework. In principle, a new level of activity and corresponding prices can be defined for an economy, a level of activity that takes account of constraints of the new kind. For a number of reasons, explained at length in the literature, however, it may be an exercise of considerable complication and somewhat little practical return to do this. Further, the institutional structure to convert the prices of "environmental capabilities" needs to be envisioned and be in place. Such a complete exercise in material balancing may at best be better attempted at local and disaggregated level. However, models of the mass balance-general equilibrium genre of Ayres and Kneese (1969), Maler (1974) and Perrings (1987) have yielded important results.

Simultaneously, alternate modes of thinking were also emerging in which environmental resources were not treated just like any other normal economic goods and services but evolution of institutions, methodological pluralism of the discipline and limitations of the prevalent neoclassical solution were identified and highlighted. While contributions of Polyani (1957) and Kapp (1972) made the beginning, Georgescu-Rogen (1971) and Herman Daly (1974) and Daly and Cobb (1989) consolidated it further, which became popular as ecological economics. In this process, many agricultural economists concerned with the biophysical dimension of agricultural productivity and growth also made their contributions, especially on issues such as soil-erosion, water logging, mono cropping, deforestation and land use planning.

There evidently exists unresolved issues at the frontiers of ecology and economics which offer interesting vistas for theoreticians to delve into. The answer to these questions must be sought at disaggregated eco-system levels to carry conviction particularly in a developing country scenario where competing demands on resources are persistent and often of a compelling kind. This is all the more significant in the context of developing countries where development shall always occupy center-stage and research questions inevitably get associated with policy issues. We need to address location specific and time related environment-development linkages and provide answers to the questions of trade-offs and synergies between environment and development. This can be done using alternative frameworks and methodologies.

As a matter of fact, economics has never been a monolithic discipline and this principle holds true in dealing with environmental and resource economics. As the issues of environment are diverse and wide ranging, economic approaches to study them are multi-dimensional. There has been a wider acceptance of Environmental and Natural Resources Economics in the teaching departments of Economics, Business Economics and Business Management in India. The response for this course has been quite encouraging from under graduates and graduate students of various colleges and university departments of Economics, Business Economics and allied areas in India. Invariably, a course in Environmental/ Natural Resources Economics is either compulsory or optional for these students in this subject in the country. During series of interactions with students and teachers of Environment and Natural Resource Economics in different colleges and universities, I got the opportunity to learn that while the courses are well crafted, readings are of good quality, availability of appropriate text books still remains a problem. This is truer in the case of countryside universities and colleges, where relevant journals and quality textbooks are either absent or are in acute scarcity. Needless to say, there is a big demand for a standard book on Environmental and Natural Resources Economics to cater to the needs of the students. The collection of some relevant articles in this volume is an attempt to cater to those long felt needs by the students and the teachers.

Most of the syllabi in environmental and ecological economics (I am using this word interchangeably as there hardly exists a difference between the two in this part of the world, however I am well aware of the differences in the two approaches in Europe and North America!) in Indian universities and colleges covers topics like valuation of environmental change (malign / benign), market based instruments for environmental management and conservation initiatives, extended benefit cost analysis, environmental and natural resources accounting, trade and environment, corporate environmental management and linkages of poverty and environmental degradation. In this volume, each topic has been given a representation and basic but analytically sound articles have been included.

The articles selected are mixed-fresh and already published by some of the acknowledged experts around the world. However this book will be quite useful for researchers and practitioners as well.

In chapter 2, Mohan Munasinghe, in a very innovative way, introduces the concept of sustainable economics called sustainomics. The whole concept has been approached as a 'trans-disciplinary, integrative, balanced, heuristic and practical meta-framework for making development more sustainable'. Being characterised as trans-disciplinary and integrative, the environmental, social and economic criteria for sustainability play an important role in the sustainomics framework. The environmental sustainability comes into play because of the concerns relating to resource degradation, pollution and loss of biodiversity, and hence overall viability and health of ecological systems (defined in terms of a comprehensive, multi scale, dynamic, hierarchical measure of resilience). Social sustainability deals with reduction of the vulnerability and maintenance of the health of social and cultural systems. Enhancing human capital through education, strengthening social values and institutions are important in this framework such that a social system develops within it a dynamic ability to adapt to change across a range of spatial and temporal scale, rather than the conservation of some 'ideal' static state. The economic sustainability refers to maximization of the flow of income that can be generated while at least maintaining the stock of assets (or capital), which yield this income flow. Sustainomics attempts to use two approaches to yield consistent and complementary results *i.e.* to provide integrated and balanced treatment of economic, social and environmental viewpoints. Firstly, when material growth is the main objective and uncertainty is not a serious problem, then focus may be to optimize economic output subject to constraints that ensure social and environmental sustainability. Secondly, an alternative objective could be the sustainability of environment. In this case emphasis tend to be on paths which are economically, socially and environmentally durable or resilient, but not necessarily growth optimizing. Another important dimension of sustainomics is that perspective of poverty and equity is extended from economic to social and environmental issues. The chapter also speaks about cost-benefit analysis and multi-criteria analysis as tools for analyzing sustainable development issues. In other words, sustainomics helps to find practical social and natural resource management options that facilitate sustainable development. A few case studies relating to energy problem have also been explained from the perspective of sustainomics.

In Chapter 3, a model called actor-network has been studied. Peter Soderbaum emphasises that models, theories and conceptual frameworks of the social sciences should not exclusively be seen as a matter of science but also of ideology. Ideology here has been defined as 'ideas about means and means'. From this perspective of ideology, the chapter argues about search for microeconomics that will be helpful in dealing with problems of environment and development. An important issue that this chapter tries to look at is whether other neoclassical economics is a useful conceptual framework to understand and illuminate issues relating to environmental degradation. It has been argued that neoclassical microeconomics is neither optimal nor or gives satisfying explanation for devising sustainable response options. Some other theoretical perspective such as institutions help in better understanding of environmental management. One important aspect to be noted is that it does not attempt to portray a model, which would do the same thing in a better manner, compared to the neoclassical approach. Rather it proposes a conceptual framework that partly focuses on different phenomena and which can do a different job from that usually attributed to neoclassical economics. The most important difference in the premises from that of neoclassical economics is that the model proposed here assumes a being to be a 'political economic person' rather than an Economic Man. This means that the 'actor' does not take decision only on the basis of market signals (which signifies that he is not only an economic man) but also rationality in terms of 'ideological orientation.' Hence, this being's relationships are both of market kind and non-market kind. In this set up the efficiency has been defined in a broader 'holistic' view, which covers a disaggregated idea of resources and impacts (both monetary and non monetary in both static and a dynamic sense of time interval) as well as an analysis in terms of patterns and profiles. An important non-monetary impact emphasised in the model is the impact on environment.

Clive Spash in Chapter 4, first gives the glimpe of the conventional valuation techniques and then throws light on cost benefit analysis. Calculating cost and benefit of environmental change is a main focal point of environmental economics. The basis of cost and benefit analysis (CBA) lies in microeconomic welfare theory. Measures of economic costs are based on opportunity costs. This chapter critically reviews a range of methods employed for CBA aspects to environmental issues. The objects of valuation that have referred to in the chapter are ecosystem functions, aesthetics, biodiversity, cultural and historical features and human health with an overview of these latter aspects. The various methods for deriving monetary valuation of non-market environmental impacts have also been surveyed. These methods are Production Function Approach (PFA), Travel Cost Method (TCM), Hedonic Price (HP), Contingent Valuation Method (CVM), Deliberate Monetary Valuation (DVM) and Benefit Transfer (BT). The first three methods are based on revealed preference of the agents, the next three are based on stated preference and the last *i.e.* BT is based on assumed preference. PFA and HP are evaluated on both direct market and implied behaviour of the agents, TCM is based on implied behaviour of the agents and all others, are based on hypothetical situation. The chapter also explains the approach *i.e.* how to get the required values (*e.g.* structural questionnaire) and operational requirements for these methods.

Chapter 5 by M N Murty deals with one of the most critical and widely used concept-economic instruments for environmental management. Use of economic instruments for pollution management and other spheres of environmental management including conserving biodiversity and ecosystems management have been popular among the policy and decision makers. Policy makers prefer the economic instruments to approaches followed in the past mostly in the domain of legal and regulatory approaches known as command and control policies. M N Murty in the paper on 'Economic Instruments for Environmental management in India' first explains the genesis of economic instruments (or interchangeably used for market based instruments) and then categorises the instruments into quantity-based and quality-based instruments. Although the focus of the instruments in this chapter is water pollution, yet the essence of the theory holds true for any type of pollution. Murty also does an assessment of existing environmental regimes in India, which are basically command and control policies. The article discusses the scope of economic instruments and shows that it is cost effective to adopt the economic instruments with respect to command and control policies for the same environmental standards. Further, the feasibility of mixed instruments and other institutional arrangements have also been discussed in Indian context.

Chapter 6 and 7 focus on the concept, methodology and examples of environmental and natural resources accounting. Chapter 6 by Peter Bartelmus, Carsten Stahmer and Jan van Tongeren deals with an integrated environmental and economic accounting framework for a SNA satellite system with the prime objective to account for the environmental damage and natural resource constraint in the process of measuring welfare. Explicitly, the objectives of the accounting framework presented in the chapter are: segregation and elaboration of all environment related flows and stocks of assets of traditional accounts, linkage of physical resource accounting with monetary environmental accounting and balance sheets, assessments of environmental costs and benefits, accounting for the maintenance of tangible wealth and elaboration and measurement of indicators of environmentally adjusted income and product. The scheme of accounting emphasises on implication of the environment for production, value added, final and intermediate demand and tangible wealth. Hence the framework does not present complete accounts for all institutional sectors. Transaction related to income distribution and those concerning intangible assets, including exploitation rights and financial assets, have also been excluded. It explains welfare-oriented measures of the economic use of the environment. The chapter compares between traditional and environmental accounting of a few important macro aggregates of in imaginary. Kirk Hamilton in chapter 7 brings into notice that market failure and policy failure both contributes to degradation of environment. It argues that policy signals provided by conventional measure of economic indicators, such as GNP are inadequate, as they do not take account of environmental degradation. However, from the point of view of sustainable development, accounting for the environmental damage is very important. The chapter explains various environmental accounts, such as Adjusted National Accounting Aggregates, Natural Resource Accounts, Resource and Pollutant Flow Accounts and Environmental Expenditure Accounts. The chapter also explains theoretical basis for environmental accounts. The theory would speak about how natural resource and pollution coupled with some other assets can be included in environmental accounting. The chapter also gives a brief note on sustainability indicators with a special emphasis on genuine saving indicators. A few selected empirical studies on genuine saving have also been surveyed in the chapter. The chapter also discusses policy issues spanning economics, ecology and social sciences with a special emphasis on use of genuine saving indicator, to the extent that it provides a rough measure of whether an economy is on a sustainable path. The next question concerns the range of policies that can influence this indicator.

Chapters 8, 9, 10 and 11 cover a wide range of issues in the domain of trade and environment. Chapter 8 by Kym Anderson deals with whether opening up of trade leads; to increase (or decrease) of welfare of a nation when the production of a good creates pollution and not consumption. The answer varies depending whether the good has been imported or is being exported. For a small country, this chapter demonstrates that if the country imports the product, then it improves the welfare of the small country. If the good is exported then due to increase in production, subsequent degradation of environment increases. The costs of degradation may not be outweighed by increased welfare due to exports. However, if a pollution tax is imposed, welfare increases in both the cases. If on the other hand, pollution is created due to consumption, without pollution tax imports may not lead to increase in welfare, but welfare increases if the commodity is being exported. In case of pollution tax, gain in welfare takes place in both the cases. Pollution tax has been shown to be the optimal instrument compared to others, in the sense that other instruments lead to increase in welfare less than what can be achieved through pollution tax.

In case of large country too, pollution tax remains optimal. This chapter shows that if the source of pollution is the process used in the production, and pollution tax is used, rather than consumption or production per se, gains in welfare is greater than when the source of pollution is either production or consumption. In other words the crux of the chapter is that optimal environmental policy such as pollution tax is more efficient than any other trade policy and infact opening up with optimal environmental policy leads to gain in welfare of an economy. In a survey (chapter 9) Karl W. Steininger looks at the links between trade and environment. It focuses on the studies in a general equilibrium framework as well as empirical work that have looked at the links between the two. It introduces how trade theories such as Hecksher-Ohlin, Factor Price Equalisation theorem and Income Distribution theory have been extended to link issues relating to the protection of environment. Also, theoretical studies concerned with issue specific extensions of the trade theories have too, been surveyed. The survey of empirical works reveal that environmental control costs have statistically insignificant effect on the pattern of trade. It strengthens the view about the limitations of econometric works that draw wrong conclusions. It also gives a brief introduction to the Computable General Equilibrium analysis with a special reference to the linkage between trade and environment. Michael Rauscher in Chapter 10 gives a brief introduction to various issues relating to trade and environment nexus. It provides a brief description of how environmental policies affect the pattern of trade. [It brings into notice that empirically environmental regulation has been found to have insignificant effect on trade, as global pollution becomes increasingly severe.] This means that if the environmental standards are tightened in the future, the relation between trade and environment may become increasingly important. This chapter introduces the issue of using environmental policies to achieve trade-policy objectives and the chapter discusses interventions in trade that are aimed at achieving environmental goals. The issue of interest group lobbying for environmental protection has also been discussed which might lead to environmental protectionism. In this context, the author points out two different kinds of environmental preferences, *true endowment i.e.* when preference for environmental protection is based on the fact that environmental goods are scarce and *de facto endowment i.e.* preference for environmental protection created by lobbying or other political process. Finally, it deals with international agreements on trade and environment, with a special reference to the importance of GATT and WTO in international trade and environment.

Kerry Krutilla in Chapter 11, reviews partial equilibrium models that look at impact of trade liberalisation on the environment and hence, determine the structure optimal environmental policy for an open economy. It emphasises on the optimal environmental policy when pollution is local and how economic adjustment can be carried out in an open economy to determine the welfare of the economy. The issue has been looked at for both large and small country, for both production and consumption, externality of pollution and the country being a net exporter and a net importer. This chapter also deals with the case where pollution crosses boundary of the domestic economy and then how trade and environment linkage can be explored.

Chapter 12 touches upon the status, drivers and tools of environmental management at the business firm level. Internalising the environmental concerns into the decisions of business corporations is very important if the economy and in turn business has to sustain in the long run. Environmental management practices can be adopted through tools like environmental accounting (firm level), auditing, reporting and other measures which are part of ISO14001. This chapter discusses these tools briefly. In adopting these tools, firm might face the situation of trade off more often than that of win-win. Then the question comes what is the incentive for the firms who invariably follow the signal of profit and market share? What are the drivers of EM especially in Indian context? How have been the experiences with environmental management practices in India been so far? This chapter attempts to address. All these qquestions.

Anantha K Duraiappah in chapter 13, comprehensively explores the linkages between poverty and environmental degradation especially in context of developing countries. One school of thought argues that poverty directly leads to environmental degradation as the poor people use natural resources without taking measures for the nature to re-grow. This chapter looks at a complex set of **relationship between** poverty and environment. It emphasises on four kinds of relationships and different policy implications for mitigating poverty and environmental degradation. These relationships are: exogenous poverty causing environmental degradation; power, wealth and greed causing environmental degradation; institutional failure causing environmental degradation, market failure causing environmental degradation, environmental degradation causing poverty [a feed back in terms of environmental degradation causing poverty it has been named endogenous poverty and environmental poverty causing environmental degradation.] Taking these relationships in view, the chapter surveys some studies available in the literature. The issues that are selected for the surveys are deforestation, land degradation, water, and air pollution. The central conclusion arrived at from the survey is that the poor do not initially or indirectly lead to degradation of environment, and their activities leading to degradation is dependent upon other groups' activities, and that powerful and wealthy degrade the environment. The most important reasons for this are market and institutional failure, the chapter reflects.

References

Ayres, R. U. and Kneese, A. V. (1969), Production, Consumption and Externalities, *American Economic Review,* Vol. 59, No 3.

Boulding, K. E. (1978), *Ecodynamics.* Sage, Beverly Hill, CA.

Clark, W. C. (1976), *Mathematical Bioeconomics: The Optimal Management of Renewable Resources*, New York, John & Wiley Sons.

Daly, H. E. (1974), The Economics of the Steady State, *American Economic Review.*

Daly, H.E. and J. B. Cobb (1989) *For the Common Good*, Beacon Press, Boston.

Georgescu-Rogen, N. (1971) *The Entropy Law and Economic Processes*, Harvard University Press, Cambridge.

Hotelling, H. (1939), The Economics of Exhaustible Resources, *Journal of Political Economy, 39.*

Kapp, K.W. (1972). Environmental Disruption and Social Costs, in *Political Economy of the Environment, Hawthorne, N.Y.: Mouton.*

Maler, K. G. (1974) *Environmental Economics, A Theoretical Enquiry.* Baltimore: Johns Hopkins University Press.

Perrings, C. (1987), *Economy and Environment*, New York: Cambridge University Press.

Pigou, A. C. (1920), *The Economics of Welfare.* London: Macmillan.

Polyani, K [1944] (1957), *The Great Transformation.* Reprint. Boston: Beacon.

2

The Sustainomics Transdisciplinary Meta-framework for Marking Development More Sustainable: Applications to Energy Issues

Mohan Munasinghe

2.1 BASIC FRAMEWORK

World decision-makers are looking for new solutions to many critical problems, including traditional development issues (such as economic stagnation, persistent poverty, hunger, malnutrition, and illness), as well as newer challenges (such as worsening environmental degradation and accelerating globalization). One key approach that has received growing attention is based on the concept of sustainable development or 'development which lasts'. Following the 1992 Earth Summit in Rio de Janeiro and the adoption of the United Nations' Agenda 21, sustainable development has become well accepted worldwide (WCED, 1987; UN, 1993).

Although no universally acceptable practical definition of sustainable development exists as yet, the concept has evolved to encompass three major points of view: economic, social, and environmental, as represented by the triangle in Figure 2.1a (see for example, Munasinghe, 1993). Each viewpoint corresponds to a domain (and system) that has its own distinct driving forces and objectives. The economy is geared mainly towards improving human welfare, primarily through increases in the consumption of goods and services. The environmental domain focuses on protection of the integrity and resilience of ecological systems. The social domain emphasizes the enrichment of human relationships and achievement of individual and group aspirations.

Meanwhile, energy has emerged as one of the key resources whose use affects the economic, social and environmental dimensions of sustainable development. First, it has long been perceived as a major driving force underlying economic progress. Second, energy production and use are strongly interlinked with the environment. Third, energy is a basic human need, which significantly affects social well-being. In recent times, growing energy demand has also become associated with global climate change, which poses an unprecedented challenge to humanity. The wide-ranging potential impacts of energy production and consumption on sustainable development suggest that the linkages between these two topics need to be critically analysed. Accordingly, this paper sketches out a transdisciplinary meta-framework (named sustainomics) and seeks to apply it to the nexus of sustainable development and energy (including climate change).

Given the lack of a specific approach or framework that attempts to define, analyse, and implement sustainable development, Munasinghe (1993, 1994) proposed the term sustainomics to describe 'a transdisciplinary, integrative, comprehensive, balanced, heuristic and practical meta-framework for making development more sustainable.' The multiplicity and complexity of issues involved cannot be covered by a single discipline. Hitherto, multidisciplinary approaches involving teams of specialists from different disciplines have been applied to sustainable development issues. A further step has been taken through interdisciplinary work, which seeks to break down the barriers among various disciplines. However, what is now required is a truly transdisciplinary metaframework, which would weave the knowledge from existing disciplines into new concepts and methods that could address the many facets of sustainable development— from concept to actual practice. Thus, sustainomics would provide a comprehensive and eclectic knowledge base to support sustainable development efforts—see Figure 2.1b.

The sustainomics approach encompasses recent initiatives on a 'sustainability transition' and 'sustainability science', and goes even further in seeking to synthesize a 'science of sustainable development', which integrates knowledge from both the sustainability and development domains (Clark, 2000; Parris and Kates, 2001; Tellus Institute, 2001). Such a synthesis will need to draw on core disciplines such as ecology, economics, and sociology, as well as anthropology, botany, chemistry, demography, ethics, geography, law, philosophy, physics, psychology, zoology, etc. Technological skills such as engineering, biotechnology (*e.g.* to enhance food production), and information technology (*e.g.* to

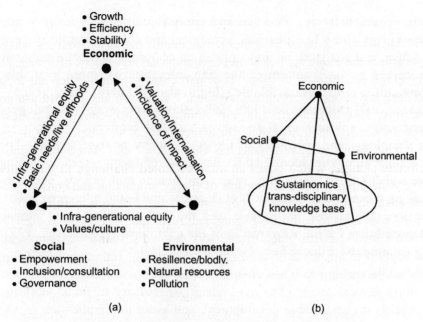

- Growth
- Efficiency
- Stability
Economic

Infra-generational equity
Basic needs/live effhoods

Valuation/Internalisation
Incidence of impact

- Infra-generational equity
- Values/culture

Social
- Empowerment
- Inclusion/consultation
- Governance

Environmental
- Resillence/blodlv.
- Natural resources
- Pollution

Economic

Social

Environmental

Sustainomics
trans-disciplinary
knowledge base

(a)   (b)

Fig. 2.1: (a) Elements of Sustainable Development.
(b) Sustainable Development Triangle Supported by
the Sustainomics Framework.

Source: Adapted from Munasinghe [1993, 1994]

improve the efficiency of natural resource use), also play a key role. Methods that bridge the economy-society-environment interfaces are especially important. For example, environmental and resource economics attempts to incorporate environmental considerations into traditional neoclassical economic analysis (Freeman, 1993; Teitenberg, 1992). The growing field of ecological economics goes further in combining ecological and economic methods to address environmental problems, and emphasizes the importance of key concepts like the scale of economic activities (for a good introduction, see Costanza *et al.*, 1997). Newer areas related to ecological science, such as conservation ecology, ecosystem management and political ecology, have led to alternative approaches to the problems of sustainability, including crucial concepts like system resilience, and integrated analysis of ecosystems and human actors (Holling, 1992). Recent papers in sociology have explored ideas about the integrative glue that binds societies together, while drawing attention to the concept of social capital and the importance of social inclusion (Putnam, 1993). The literature on energetics and energy economics has focussed on the relevance of physical laws, such as the first and second laws of thermodynamics (covering mass/energy balance and entropy, respectively). This research has yielded valuable insights into how energy flows link physical, ecological and socioeconomic systems together, and analysed the limits placed on ecological and socioeconomic processes by laws governing the transformation of 'more available' (low entropy) to 'less available' (high entropy) energy (Georgescu—Roegen, 1971; Munasinghe, 1990; Hall, 1995). Recent work on sociological economics, environmental sociology, cultural economics, economics of sociology, and sociology of the environment is also relevant. The literature on environmental ethics has explored many issues, including the weights to be attached to values and human motivations, decision-making processes, consequences of decisions, intra- and inter-generational equity, the 'rights' of animals and the rest of nature, and human responsibility for the stewardship of the environment (Andersen, 1993; Environmental Ethics; Sen, 1987; Westra, 1994).

While seeking to build on such earlier work, sustainomics projects a more neutral image. The neologism is necessary to focus attention explicitly on sustainable development, and avoid the implication of any disciplinary bias or hegemony. For example, both biology and sociology can provide important insights into human behaviour, which challenge the 'rational actor' assumptions of neoclassical economics. Thus, recent studies seek to explain phenomena such as hyperbolic discounting (versus the more conventional exponential discounting), reciprocity, and altruistic responses (as opposed to selfish, individualistic behaviour) (Gintis, 2000; Robson, 2001). In the same vein, Siebhuner (2000) has sought to define 'homo sustinens' as a moral, cooperative individual with social, emotional, and nature-related skills, as opposed to the conventional 'homo economicus' motivated primarily by economic self interest and competitive instincts. The substantive trans-disciplinary framework underlying sustainomics is the precursor of a more rigorous 'science of sustainable development'. The approach should lead to the balanced and consistent treatment of the economic, social and environmental dimensions of sustainable development (as well as other relevant disciplines and paradigms). Balance is also needed in the relative emphasis placed on traditional development versus sustainability. For example, much of the mainstream literature on sustainable development which originates in the North tends to focus on pollution, the unsustainability of growth, and population increase. These ideas have far less resonance in the South, whose priorities include continuing development, consumption and growth, poverty alleviation, and equity.

While sustainable development itself involves every aspect of human activity, including complex interactions among socioeconomic, ecological, and physical systems, Many disciplines contribute to the sustainomics framework, The scope of analysis needs to extend from the global to the local scale, cover time spans extending to centuries (for example, in the case of climate change), and deal with problems of uncertainty, irreversibility, and non-linearity. The sustainomics framework seeks to establish an overarching design for analysis and policy guidance, while the constituent components (or disciplines) provide the 'reductionist' building blocks and foundation. The heuristic element underlines the need for continuous rethinking based on new research, empirical findings, and current best practice, because reality is more complex than our models, our understanding is incomplete, and we have no consensus on the subject. Furthermore, the precise definition of sustainable development remains an elusive (and perhaps unreachable) goal. Thus, a less ambitious strategy that merely seeks to make development more sustainable might offer greater promise. Such an incremental (or gradient-based) method is more practical, because many unsustainable activities may be easier to recognize and eliminate. In particular, it will help us avoid sudden catastrophic ('cliff edge') outcomes.

This paper identifies some of the key constituent elements of sustainomics and how they might fit together. It also illustrates some of these concepts, by applying them to case studies involving energy problems (the theme of this special issue of *IJSD*) across the full range of spatial scales—at the global-transnational, national-economy, subnational-sectoral, and local-project levels. The current state of knowledge is inadequate to provide a comprehensive definition of sustainomics. Furthermore, sustainomics must provide a heuristic, dynamically evolving framework, in order to address rapidly changing sustainable development issues. Therefore, the intention here is to sketch out several preliminary ideas that would serve as a starting point, thereby stimulating discussion and encouraging further contributions that are needed to flesh out the initial framework.

2.2 SOME ELEMENTS OF SUSTAINOMICS

Current approaches to sustainable development draw on the experience of several decades of development efforts. Historically, the development of the industrialized world focussed on material production. Not surprisingly, most industrialized and developing nations have pursued the economic goal of increasing output and growth during the twentieth century. Thus, the traditional approach to development was strongly associated with economic growth, but has important social dimensions as well (see the section on poverty and equity, below).

By the early 1960s the large and growing numbers of poor in the developing world, and the lack of 'trickle-down' benefits to them, resulted in greater efforts to improve income distribution directly. The development paradigm shifted towards equitable growth, where social (distributional) objectives, especially poverty alleviation, were recognized to be as important as economic efficiency, and distinct from the latter (see the section on poverty and equity, below).

Protection of the environment has now become the third major objective of sustainable development. By the early 1980s, a large body of evidence had accumulated that environmental degradation was a major barrier to development, and new proactive safeguards were introduced (such as the environmental assessments).

Broadly speaking, sustainable development may be described as a process for improving the range of opportunities that will enable individual human beings and communities to achieve their aspirations and full potential over a sustained period of time, while maintaining the resilience of economic, social and environmental systems (Munasinghe 1994). In other words, sustainable development requires increases both in adaptive capacity and in opportunities for improvement of economic, social and ecological systems (Gunderson and Holling 2001). Improving adaptive capacity will increase resilience and sustainability. Expanding the set of opportunities for improvement will give rise to development. Heuristic behaviour of individual organisms and systems facilitates learning, the testing of new processes, adaptation, and improvement. Adapting this general concept, a more focussed and practical approach towards making development more sustainable would seek continuing improvements in the present quality of life at a lower intensity of resource use, thereby leaving for future generations an undiminished stock of productive assets (*i.e.*, manufactured, natural, and social capital) that will enhance opportunities for improving their quality of life.

2.2.1 Economic Aspects

Economic progress is often evaluated in terms of welfare (or utility)—measured as willingness to pay for goods and services consumed. Thus, many economic policies typically seek to enhance income, and induce more efficient production and consumption of (mainly marketed) goods and services. The stability of prices and employment are among other important objectives. At the same time, the equation of welfare with monetary income and consumption has been challenged for many years. For example, Buddhist philosophy (over 2500 years old) still stresses that contentment is not synonymous with material consumption (Ven. Narada, 1988). More recently, Maslow (1970) and others have identified hierarchies of needs that provide psychic satisfaction, beyond mere goods and services.

The degree of economic efficiency is measured in relation to the ideal of Pareto optimality, which encourages actions that will improve the welfare of at least one individual without worsening the situation of anyone else. The idealized, perfectly competitive economy is an important (Pareto optimal) benchmark, where (efficient) market prices play a key role in both allocating productive resources to maximize output, and ensuring optimal consumption choices which maximize consumer utility. If significant economic distortions are present, appropriate shadow prices need to be used. The well known cost-benefit criterion accepts all projects whose net benefits are positive (*i.e.* aggregate benefits exceed costs) (Munasinghe, 1993). It is based on the weaker 'quasi' Pareto condition, which assumes that such net benefits could be redistributed from the potential gainers to the losers, so that no one is worse off than before. More generally, interpersonal comparisons of (monetized) welfare are fraught with difficulty—both within and across nations, and over time (*e.g.* the value of human life).

2.2.1.1 *Economic Sustainability*

The modern concept underlying economic sustainability seeks to maximize the flow of income that could be generated while at least maintaining the stock of assets (or capital) which yield these beneficial outputs (Solow, 1986; Maler, 1990). This approach is based on the pioneering work of Lindahl and Hicks. For example, Hicks (1946) implies that people's maximum sustainable consumption is 'the amount that they can consume without impoverishing themselves'. Much earlier Fisher (1906) had defined *capital* as 'a stock of instruments existing at an instant of time', and *income* as 'a stream of services flowing from this stock of wealth'. Economic efficiency continues to play a key role—in ensuring both efficient allocation of resources in production, and efficient consumption choices that maximize utility. Problems of interpretation arise in identifying the kinds of capital to be maintained (for example, manufactured, natural, and human resource stocks, as well as social capital have been identified) and their substitutability (see next section). Often, it is difficult to value these assets and the services they provide, particularly in the case of ecological and social resources (Munasinghe, 1993). Even key economic assets may be overlooked, for example, in informal or subsistence economies where non-market based transactions are important. The issues of uncertainty, irreversibility, and catastrophic collapse pose additional difficulties in determining dynamically efficient development paths (Pearce and Turner, 1990). Many commonly used microeconomic approaches rely heavily on marginal analysis based on small perturbations (*e.g.* comparing incremental costs and benefits of economic activities). From the viewpoint of resilience theory (discussed below), this type of system soon returns to its dominant stable equilibrium and thus there is little risk of instability. Such methods assume smoothly changing variables and are therefore rather inappropriate for analysing large changes, discontinuous phenomena, and sudden transitions among multiple equilibria. More recent work (especially at the cutting edge of the economics-ecology interface) has begun to explore the behaviour of large, non-linear, dynamic, and chaotic systems, as well as newer concepts like system vulnerability and resilience.

2.2.2 Environmental Aspects

Development in the environmental sense is a rather recent concern relating to the need to manage scarce natural resources in a prudent manner—because human welfare ultimately depends on ecological services. Ignoring safe ecological limits will increase the risk of undermining long-run prospects for development. Dasgupta and Maler (1997) point out that until the 1990s, the mainstream development literature hardly mentioned the topic of environment (see for example, Stern, 1989; Chenery and Srinivasan, 1988, 1989; and Dreze and Sen, 1990). An even more recent review paper on economic growth in the well-known *Journal of Economic Literature* mentions the role of natural resources only in the passing (Temple, 1999). Examples of the growing literature on the theme of environment and sustainable development include books by Faucheux *et al.* (1996) describing models of sustainable development, and Munasinghe *et al.* (2001) explicitly addressing the links between growth and environment.

2.2.2.1 Environmental Sustainability

The environmental interpretation of sustainability focuses on the overall viability and health of ecological systems—defined in terms of a comprehensive, multiscale, dynamic, hierarchical measure of resilience, vigour and organization (Costanza, 2000). The classic definition of resilience was provided by Holling (1973) in terms of the ability of an ecosystem to persist despite external shocks. Resilience is determined by the amount of change or disruption that will cause an ecosystem to switch from one system state to another. An ecosystem state is defined by its internal structure and set of mutually reenforcing processes. Petersen *et al.* (1998) argue that the resilience of a given ecosystem depends on the continuity of related ecological processes at both larger and smaller spatial scales (see Box 2.1). Further discussion of resilience may be found in Pimm (1991), and Ludwig *et al.* (1997). Vigour is associated with the primary productivity of an ecosystem. It is analogous to output and growth as an indicator of dynamism in an economic system. Organization depends on both complexity and structure in an ecological or biological system. For example, a multicellular organism like a human being is more highly organized (having more diverse subcomponents and interconnections among them), than a single celled amoeba. Higher states of organization imply lower levels of entropy. Thus, the second law of thermodynamics requires that the sustainability of more complex organisms depends on the use of low entropy energy derived from their environment, which is returned as (less useful) high entropy energy. The ultimate source of this energy is solar radiation.

Box 2.1: Spatial and Temporal Aspects of Sustainability

An operationally useful concept of sustainability must refer to the persistence, viability and resilience of organic or biological systems, over their 'normal' lifespan (see the main text for a discussion of resilience). In this ecological context, sustainability is linked with both spatial and temporal scales, as shown in Figure 2.2. The X axis indicates lifetime in years and the Y axis shows linear size (both in logarithmic scale). The central O represents an individual human being-having a longevity and size of the order of 100 years and 1.5 metres, respectively. The diagonal band shows the expected or 'normal' range of lifespans for a nested hierarchy of living systems (both ecological and social), starting with single cells and culminating in the planetary ecosystem. The bandwidth accommodates the variability in organisms as well as longevity.

Environmental changes that reduce lifespans below the normal range imply that external conditions have made the systems under consideration unsustainable. In short, the regime above and to the left of the normal range denotes premature death or collapse. At the same time, it is unrealistic to expect any system to last forever. Indeed, each subsystem of a larger system (such as single cells within a multi-cellular organism) generally has a shorter lifespan than the larger system itself. If subsystem life spans increase too much,

---

(contd.)

Box 2.1 (contd.)

the system above it is likely to lose its plasticity and become 'brittle'—as indicated by the region below and to the right of the normal range (Holling, 1973). In other words, it is the timely death and replacement of subsystems that facilitate successful adaptation, resilience and evolution of larger systems.

Gunderson and Holling (2001) use the term 'panarchy' to denote such a nested hierarchy of systems and their adaptive cycles across scales. A system at a given level is able to operate in its stable (sustainable) mode, because it is protected by the slower and more conservative changes in the super-system above it, while being simultaneously invigorated and energized by the faster cycles taking place in the sub-systems below it. In brief, both conservation and continuity from above, and innovation and change from below, play a useful role in the panarchy.

We may argue that sustainability requires biological systems to be able to enjoy a normal life span and function normally, within the range indicated in Figure 2.2. Thus, leftward movements would be especially undesirable. For example, the horizontal arrow might represent a case of infant death-indicating an unacceptable deterioration in human health and living conditions. In this specific case, extended longevity involving a greater than normal lifespan would not be a matter for particular concern. On the practical side, forecasting up to a timescale of even several hundred years is rather imprecise. Thus, it is important to improve the accuracy of scientific models and data, in order to make very long-term predictions of sustainability (or its absence) more convincing-especially in the context of persuading decision makers to spend large sums of money to reduce unsustainability. One way of dealing with uncertainty, especially if the potential risk is large, relies on a precautionary approach—*i.e.* avoiding unsustainable behaviour using low cost measures, while studying the issue more carefully.

To conclude, sustainable development of ecological systems requires both adaptive capacity and opportunities for improvement. Improving adaptive capacity will increase resilience and sustainability. Expanding the set of opportunities for system improvement will give rise to development. Heuristic system behaviour facilitates learning, the testing of new processes, adaptation, and improvement.

In this context, natural resource degradation, pollution and loss of biodiversity are detrimental because they increase vulnerability, undermine system health, and reduce resilience (Perrings and Opschoor, 1994;

Munasinghe and Shearer, 1995). The notion of a safe threshold (and the related concept of carrying capacity) are important—often to avoid catastrophic ecosystem collapse (Holling, 1986). It is useful to also think of sustainability in terms of the normal functioning and longevity of a nested hierarchy of ecological and socioeconomic systems, ordered according to scale—*e.g.* a human community would consist of many individuals, who are themselves composed of a large number of cells (see Box 2.1 for details). Gunderson and Holling (2001) use the term 'panarchy' to denote such a hierarchy of systems and their adaptive cycles across scales. A system at a given level is able to operate in its stable (sustainable) mode, because it is protected by the slower and more conservative changes in the super-system above it, while being simultaneously invigorated and energized by the faster cycles taking place in the sub-systems below it. In brief, both conservation and continuity from above, and innovation and change from below, are integral to the panarchy-based approach, helping to resolve the apparent paradox between the need for stability as well as change.

Sustainable development is not necessarily synonymous with the maintenance of the ecological *status quo*. From an economic perspective, a coupled ecological socio-economic system should evolve so as to maintain a level of biodiversity that will guarantee the resilience of the ecosystems on which human consumption and production depend.

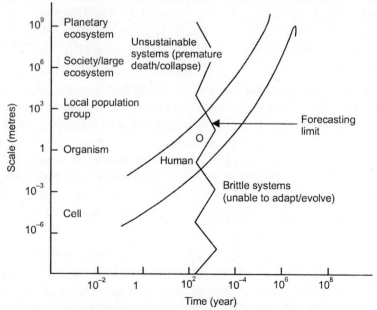

Fig. 2.2: Spatial and Temporal Norms for Sustainable Biological and Social Systems.

Sustainable development demands compensation for the opportunities foregone by future generations, because today's economic activity changes the level or composition of biodiversity in a way that will affect the flow of vital future ecological services, and narrow the options available to unborn generations. This holds true even if positive rates of economic growth indicate an increase in the instrumental (or use) values of options currently available.

2.2.3 Social Aspects

Social development usually refers to improvements in both individual well-being and the overall welfare of society (more broadly defined), that result from increases in social capital—typically, the accumulation of capacity for individuals and groups of people to work together to achieve shared objectives. The institutional component of social capital refers mainly to the formal laws as well as to traditional or informal understandings that govern behaviour, while the organizational component is embodied in the entities (both individuals and social groups) that operate within these institutional arrangements. The quantity and quality of social interactions that underlie human existence, including the level of mutual trust and extent of shared social norms, help to determine the stock of social capital. Thus social capital tends to grow with greater use and erodes through disuse, unlike economic and environmental capital, which are depreciated or depleted by use. Furthermore, some forms of social capital may be harmful (*e.g.* cooperation within criminal gangs may benefit them, but impose far greater costs on the larger community).

There is an important element of equity and poverty alleviation as well (see below). Thus, the social dimension of development includes protective strategies that reduce vulnerability, improve equity and ensure that basic needs are met. Future social development will require socio-political institutions that can adapt to meet the challenges of modernization—which often destroy traditional coping mechanisms that have evolved in the past (especially to protect disadvantaged groups).

2.2.3.1 Social Sustainability

Social sustainability is able to draw on the ideas discussed earlier regarding environmental sustainability, since habitats may be interpreted broadly to include manmade environments like cities and villages (UNEP, IUCN, and WWF, 1991). Reducing the vulnerability and maintaining the health (*i.e.* resilience, vigour and organization) of social and cultural systems, and their ability to withstand shocks, is also important (Chambers, 1989; Bohle *et al.*, 1994; Ribot *et al.*, 1996). Enhancing human capital (through education) and strengthening social values and institutions (like trust and behavioural norms) are key aspects. Weakening social values, institutions, and equity will reduce the resilience of social systems and undermine governance. Many such harmful changes occur slowly, and their long-term effects are often overlooked in socio-economic analysis. Preserving cultural diversity and cultural capital across the globe, strengthening social cohesion and networks of relationships, and reducing destructive conflicts are integral elements of this approach. An important aspect of empowerment and broader participation is subsidiarity—*i.e.* decentralization of decision making to the lowest (or most local) level at which it is still effective. In summary, for both ecological and socioeconomic systems, the emphasis is on improving system health and its dynamic ability to adapt to change across a range of spatial and temporal scales, rather than the conservation of some 'ideal' static state (see also Box 2.1).

2.2.4 Equity and Poverty

Equity and poverty are two important issues in the sustainomics framework, which have social, economic and environmental dimensions— see Figure 1.1(a). Recent worldwide statistics are compelling. Over 2.8 billion people (almost half the global population) live on less than US$2 per day, and 1.2 billion barely survive on under US$1 per day. The top 20 percentile of the world's population consumes about 83 per cent of total output, while the bottom 20 percentile consumes only 1.4 per cent. Income disparities are worsening—the per capita ratio between the richest and the poorest 20 percentile groups was 30 to 1 in 1960 and over 80 to 1 by 1995. In poor countries, up to half the children under five years of age are malnourished, whereas the corresponding figure in rich countries is less than 5 per cent.

Equity is an ethical and usually people-oriented concept with primarily social, and some economic and environmental dimensions. It focuses on the basic fairness of both the processes and outcomes of decision-making. The equity of any action may be assessed in terms of a number of generic approaches, including parity, proportionality, priority, utilitarianism, and Rawlsian distributive justice. For example, Rawls (1971) stated that 'Justice is the first virtue of social institutions, as truth is of systems of thought'. Societies normally seek to achieve equity by balancing and combining several of these criteria.

Poverty alleviation, improved income distribution and intra-generational (or spatial) equity are key aspects of economic policies seeking to increase overall human welfare (Sen, 1981, 1984). Brown (1998) points out shortcomings in utilitarianism, which underlies much of the economic approach to equity. Broadly speaking, economic efficiency provides guidance on producing and consuming goods and services more efficiently, but is unable to provide a means of choosing (from a social perspective) among various patterns of consumption that are efficient. Equity principles provide better tools for making judgements about such choices.

Social equity is also linked to sustainability, because highly skewed or unfair distributions of income and social benefits are less likely to be acceptable or lasting in the long run. Equity is likely to be strengthened by enhancing pluralism and grass-roots participation in decision-making, as well as by empowering disadvantaged groups (defined by income, gender, ethnicity, religion, caste, etc.) (Rayner and Malone, 1998). In the long term, considerations involving inter-generational equity and safeguarding the rights of future generations are key factors. In particular, the economic discount rate plays a key role with respect to both equity and efficiency aspects (Arrow *et al.*, 1995). Further details of equity-efficiency interactions that need to be reconciled within the sustainomics framework are reviewed in Box 2.2. Equity in the environmental sense has received more attention recently, because of the disproportionately greater environmental damages suffered by disadvantaged groups. In the same vein, poverty alleviation efforts (which traditionally focussed on raising monetary incomes), are being broadened to address the degraded environmental and social conditions facing the poor.

In summary, both equity and poverty have not only economic but also social and environmental dimensions and, therefore, they need to be assessed using a comprehensive set of indicators (rather than income distribution alone). From an economic policy perspective, emphasis needs to be placed on expanding employment and gainful opportunities for poor people through growth, improving access to markets, and increasing both assets and education. Social policies would focus on empowerment and inclusion, by making institutions more responsive to the poor, and removing barriers that exclude disadvantaged groups. Environmentally related measures to help poor people might seek to reduce their vulnerability to disasters and extreme weather events, crop failures, loss of employment, sickness, economic shocks, etc. Thus, an important objective of poverty alleviation is to provide poor people with assets (*e.g.* enhanced physical, human and financial resources) that will reduce their vulnerability. Such assets increase the capacity for both coping (*i.e.* making short-run changes) and adapting (*i.e.* making permanent adjustments) to external shocks (Moser, 1998). The foregoing ideas merge quite naturally with the sustainable livelihoods approach, which focuses on access to portfolios of assets (social, natural and manufactured), the capacity to withstand shocks, gainful employment, and social processes, within a community or individual oriented context.

An even broader non-anthropocentric approach to equity involves the concept of fairness in the treatment of non-human forms of life or even inanimate nature. One view asserts that humans have the responsibility of prudent 'stewardship' (or 'trusteeship') over nature, which goes beyond mere rights of usage (see for example, Brown, 1998).

Box 2.2: Interactions Between Social Equity and Economic Efficiency

Conflicts between economic efficiency and equity may arise due to assumptions about the definition, comparison and aggregation of the welfare of different individuals or nations. For example, efficiency often implies maximization of output subject to resource constraints. The common assumption is that increases in average income *per capita* will make most or all individuals better off. However, this approach can potentially result in a less equitable income distribution. Overall welfare could drop depending on how welfare is defined in relation to the distribution of income. Conversely, total welfare might increase if policies and institutions can ensure appropriate resource transfers-typically from the rich to the poor.

In the same context, aggregating and comparing welfare across different countries is a disputable issue. Gross National Product (GNP) is simply a measure of the total measurable economic output of a country, and does not represent welfare directly. Aggregating GNP across nations is not necessarily a valid measure of global welfare. However, national economic policies frequently focus more on the growth of GNP rather than its distribution, indirectly implying that additional wealth is equally valuable to rich and poor alike, or that there are mechanisms to redistribute wealth in a way that satisfies equity goals. Attempts have been made to incorporate equity considerations within a purely economic framework, by the weighting of costs and benefits so as to give preference to the poor. Although systematic procedures exist for determining such weights, often the

---

(contd.)

Box 2.2 (contd.)

element of arbitrariness in assigning weights has caused many practical problems.

At the same time, it should be recognized that all decision-making procedures do assign weights (arbitrarily or otherwise). For example, progressive personal income taxes are designed to take proportionately more from the rich. On the other hand, traditional cost-benefit analysis based on economic efficiency (which seeks to maximize net benefits) assigns the same weight of unity to all monetary costs and benefits-irrespective of income levels. More pragmatically, in most countries the tension between economic efficiency and equity is resolved by keeping the two approaches separate, *e.g.* by maintaining a balance between maximizing GNP, and establishing institutions and processes charged with redistribution, social protection, and provision of various social goods to meet basic needs. The interplay of equity and efficiency at the international level is illustrated later, in the climate change case study.

2.2.5 Integration of Economic, Social and Environmental Considerations

As a prelude to integration, it is useful to compare the concepts of ecological, social, and economic sustainability. One useful idea is that of the maintenance of the set of opportunities, as opposed to the preservation of the value of the asset base (Githinji and Perrings, 1992). In fact, if preferences and technology vary through successive generations, merely preserving a constant value of the asset base becomes less meaningful. By concentrating on the size of the opportunity set, the importance of biodiversity conservation becomes more evident, for the sustainability of an ecosystem. The preservation of biodiversity allows the system to retain resilience by protecting it from external shocks, in the same manner that preservation of the capital stock protects economic assets for future consumption. Differences emerge because under the Hicks-Lindahl income measure, a society that consumes its fixed capital without replacement is not sustainable, whereas using an ecological approach, loss of resilience implies a reduction in the self-organization of the system, but not necessarily a loss in productivity. In the case of social systems, resilience depends to a certain extent on the capacity of human societies to adapt and continue functioning in the face of stress and shocks. Thus, linkages between socio-cultural and ecological sustainability emerge through the organizational similarities between human societies and ecological systems, and the parallels between biodiversity and cultural diversity. From a longer term perspective, the concept of co-evolution of social, economic and ecological systems, within a larger, more complex adaptive system, provides useful insights regarding the harmonious integration of the various elements of sustainable development—see Figure 2.1(a) (Munasinghe, 1994; Costanza, 1997).

One may conclude that the exact definition of sustainable development paths is likely to be extremely difficult at this stage, and may be considered a long-run or ideal objective. However, a more promising and practical shorter run goal that is consistent with the sustainomics approach, is to seek strategies that might make future development prospects more sustainable. In such an approach, one key step would be to begin by eliminating the many unsustainable activities that are readily identifiable.

It is important to integrate and reconcile the economic, social and environmental aspects within a holistic and balanced sustainable development framework. Economic analysis has a special role in contemporary national policy making, since some of the most important decisions fall within the economic domain. While mainstream economics which is used for practical policy making has often ignored many crucial aspects of the environmental and social dimensions of sustainable development, there is a small but growing body of literature which seeks to address such shortcomings—see for example, recent issues of the journals *Ecological Economics* and *Conservation Ecology* (published on the internet).

Two broad approaches are relevant for integrating the economic, social and environmental dimensions of sustainable development. They are distinguished by the degree to which the concepts of *optimality* and *durability* are emphasized. While there are overlaps between the two approaches, the main thrust is somewhat different in each case. Uncertainty often plays a key role in determining which approach would be preferred. For example, relatively steady and well-ordered conditions may encourage optimizing behaviour that attempts to control and even fine-tune outcomes, whereas a subsistence farmer facing chaotic and unpredictable circumstances might opt for a more durable response that simply enhances survival prospects.

2.2.6 Optimality

The optimality-based approach has been widely used in economic analysis to generally maximize welfare (or utility), subject to the requirement that the stock of productive assets (or welfare itself) is non-decreasing in the long term This assumption is common to most sustainable economic growth models—for useful reviews, see Pezzey (1992) and Islam (2001). The essence of the approach is illustrated by the simple example of maximization of the flow of aggregate welfare (W), cumulatively discounted over infinite time (t), as represented by the expression:

$$\max \int_0^\infty W(C, Z)\, e^{-rt}\, dt$$

Here, C represents the consumption rate, Z is a set of other relevant variables, and r is the discount rate. Side constraints might be imposed to satisfy sustainability requirements—*e.g.* non-decreasing stocks of productive assets (including natural resources).

Some ecological models also optimize variables such as energy use, nutrient flow, or biomass production—giving more weight to system vigour as a measure of sustainability. In economic models, utility is often measured mainly in terms of the net benefits of economic activities, *i.e.* the benefits derived from development activities minus the costs incurred to carry out those actions (for more details about valuation, see Box 2.3 below, and Munasinghe, 1993, or Freeman, 1993). More sophisticated economic optimization approaches seek to include environmental and social variables (*e.g.* by attempting to value environmental externalities, system resilience, etc.). However, given the difficulties of quantifying and valuing many such 'non-economic' assets, the costs and benefits associated with market-based activities tend to dominate in most economic optimization models.

Basically, the optimal growth path maximizes economic output, while the sustainability requirement is met (within this framework) by ensuring non-decreasing stocks of assets (or capital). Some analysts support a 'strong sustainability' constraint, which requires the separate preservation of each category of critical asset (for example, manufactured, natural, socio-cultural and human capital), assuming that they are complements rather than substitutes. One version of this rule might correspond roughly to maximizing economic output, subject to side constraints on environmental and social variables that are deemed critical for sustainability (*e.g.* biodiversity loss or meeting the basic needs of the poor). Other researchers have argued in favour of 'weak sustainability', which seeks to maintain the aggregate monetary value of the total stock of assets, assuming that the various asset types may be valued and that there is some degree of substitutability among them (see for example, Nordhaus and Tobin, 1972).

Side constraints are often necessary because the underlying basis of economic valuation, optimization and efficient use of resources may not be easily applied to ecological objectives, such as protecting biodiversity and improving resilience, or to social goals, such as promoting equity, public participation and empowerment. Thus, such environmental and social variables cannot be easily combined into a single valued objective function with other measures of economic costs and benefits (see sections on cost-benefit and multi-criteria analysis, below). Moreover, the price system (which has time lags) might fail to anticipate reliably irreversible environmental and social harm, and non-linear system responses that could lead to catastrophic collapse. In such cases, non-economic indicators of environmental and social status would be helpful—*e.g.* area under forest cover, and incidence of conflict (see for example, Munasinghe and Shearer, 1995; Hanna and Munasinghe, 1995; UNDP, 1998; World Bank, 1998). The constraints on critical environmental and social indicators are proxies representing safe thresholds, which help to maintain the viability of those systems. In this context, techniques like multicriteria analysis may be required, to facilitate trade-offs among a variety of noncommensurable variables and objectives (see for example, Meier and Munasinghe, 1994). Risk and uncertainty will also necessitate the use of decision analysis tools (for a concise review of climate change decision-making frameworks, see Toth, 1999). Recent work has underlined the social dimension of decision science, by pointing out that risk perceptions are subjective and depend on the risk measures used, as well as other factors such as ethno-cultural background, socio-economic status, and gender (Bennet, 2000).

2.2.7 Durability

The second broad integrative approach would focus primarily on sustaining the quality of life—*e.g.* by satisfying environmental, social, and economic sustainability requirements. Such a framework favours 'durable' development paths that permit growth, but are not necessarily economically optimal. There is more willingness to trade off some economic optimality for the sake of greater safety, in order to stay within critical environmental and social limits—especially among increasingly risk-averse and vulnerable societies or individuals who face chaotic and unpredictable conditions (see the discussion on the precautionary principle in Section 2.3.1). The economic constraint might be framed in terms of maintaining consumption levels (defined broadly to include environmental services, leisure and other 'non-economic' benefits)—*i.e. per capita* consumption that never falls below some minimum level, or is non-declining. The environmental and social sustainability requirements may be expressed in terms of indicators of 'state' that seek to measure the durability or health (resilience, vigour and organization) of complex ecological and socio-economic systems. As an illustrative example, consider a simple durability index (D) for an ecosystem measured in terms of its expected lifespan (in a healthy state), as a fraction of the normal lifespan (see also Box 2.1). We might specify: $D = D(R, V, O, S)$, to indicate the dependence of durability on resilience (R), vigour (V), organization (O), and the state of the external environment (S)—especially in relation to potentially damaging shocks. There is the likelihood of further interaction here, owing to linkages between the sustainability of social and ecological systems—*e.g.* social disruption and conflict could exacerbate damage to ecosystems, and *vice versa*. For example, long-standing social norms in many traditional societies have helped to protect the environment (Colding and Folke, 1997).

Durability encourages a holistic systemic viewpoint, which is important in sustainomics analysis. The self-organizing and internal structure of ecological and socioeconomic systems makes 'the whole more durable (and valuable) than the sum of the parts'. A narrow definition of efficiency based on marginal analysis of individual components may be misleading (Schutz, 1999). For example, it is more difficult to value the integrated functional diversity in a forest ecosystem than the individual species of trees and animals. Therefore, the former is more likely to fall victim to market failure (as an externality). Furthermore, even where correct environmental shadow prices prevail, some analysts point out that cost minimization could lead to homogenization and consequent reductions in system diversity (Daly and Cobb, 1989; Perrings *et al.*, 1995). Systems analysis also helps to identify the benefits of cooperative structures and behaviour, which a more partial analysis may neglect.

The possibility of many durable paths favours simulation-based methods, including consideration of alternative world views and futures (rather than one optimal result). This approach is consonant with recent research on integrating human actors into ecological models (*Ecological Economics*, 2000). Key elements include multiple-agent modelling to account for heterogeneous behaviour, recognition of bounded rationality leading to different perceptions and biases, and more emphasis on social interactions that give rise to responses like imitation, reciprocity and comparison.

In the durability approach, constraints based on sustainability could be represented also by the approach discussed earlier, which focuses on maintaining stocks of assets. Here, the various forms of capital are viewed as a bulwark that decreases vulnerability to external shocks and reduces irreversible harm, rather than mere accumulations of assets that produce economic outputs. System resilience, vigour, organization and ability to adapt will depend dynamically on the capital endowment as well as the magnitude and rate of change of a shock.

2.2.8 Indicators

In view of the importance of asset stocks to both the optimal and durable approaches, the practical implementation of sustainomics principles will require the identification of specific economic, social and environmental indicators, at different levels of aggregation ranging from the global/ macro to local/micro, that are relevant. It is important that the indicators be comprehensive in scope, multi-dimensional in nature (where appropriate), and account for spatial differences. A wide variety of indicators are described already in the literature (Munasinghe and Shearer, 1995; UNDP, 1998; World Bank, 1998; Liverman *et al.*, 1988; Kuik and Verbruggen, 1991; Opschoor and Reijnders, 1991; Holmberg and Karlsson, 1992; Adriaanse, 1993; Alfsen and Saebo, 1993; Bergstrom, 1993; Gilbert and Feenstra, 1994; Moffat, 1994; OECD, 1994; Azar, 1996; UN, 1996; Commission on Sustainable Development (CSD), 1998; World Bank, 1997).

Measuring economic, environmental (natural), human and social capital also raises various problems. Manufactured capital may be estimated using conventional neoclassical economic analysis. As described later in the section on cost-benefit analysis, market prices are useful when economic distortions are relatively low, and shadow prices could be applied in cases where market prices are unreliable (*e.g.* Squire and van der Tak, 1975). Natural capital needs to be quantified first in terms of key physical attributes. Typically, damage to natural capital may be assessed by the level of air pollution (*e.g.* concentrations of suspended particulate, sulfur dioxide or GHGs), water pollution (*e.g.* BOD or COD), and land degradation (*e.g.* soil erosion or deforestation). Then the physical damage could be valued using a variety of techniques based on environmental and resource economics (*e.g.* Munasinghe, 1993; Freeman, 1993; Teitenberg, 1992). Human resource stocks are often measured in terms of the value of educational levels, productivity and earning potential of individuals. Social capital is the one that is most difficult to assess (Grootaert, 1998). Putnam (1993) described it as 'horizontal associations' among people, or social networks and associated behavioural norms and values, which affect the productivity of communities. A somewhat broader view was offered by Coleman (1990), who viewed social capital in terms of social structures, which facilitate the activities of agents in society—this permitted both horizontal and vertical associations (like firms). An even wider definition is implied by the institutional approach espoused by North (1990) and Olson (1982), which includes not only the mainly informal relationships implied by the earlier two views, but also the more formal frameworks provided by governments, political systems, legal and constitutional provisions, etc. Recent work has sought to distinguish between social and political capital (*i.e.* the networks of power and influence that link individuals and communities to the higher levels of decision-making).

2.2.9 Complementarity and Convergence of Optimal and Durable Approaches

National economic management often provides good examples of how the two approaches complement one another. For example, economy-wide policies involving both fiscal and monetary measures (*e.g.* taxes, subsidies, interest rates and foreign exchange rates) might be optimized on the basis of quantitative macroeconomic models. Nevertheless, decision-makers inevitably modify these economically 'optimal' policies before implementing them, to take into account other sociopolitical considerations based more on durability (such as protection of the poor, regional factors, etc.), which facilitate governance and stability. The determination of an appropriate target trajectory for future global GHG emissions (and corresponding target GHG concentration) provides another useful illustration of the interplay of the durability and optimality approaches (for details see IPCC, 1996a; Munasinghe, 1998a, and Case Study 1 below).

The practical potential for convergence of the two approaches may be realized in several ways. First, wastes ought to be generated at rates less than or equal to the assimilative capacity of the environment—for example, emissions of greenhouse gases and ozone-depleting substances into the global atmosphere. Second, renewable resources, especially if they are scarce, should be utilized at rates less than or equal to the natural rate of regeneration. Third, non-renewable resource use should be managed in relation to the substitutability between these resources and technological progress. Both wastes and natural resource input use might be minimized by moving from the linear throughput to the closed loop mode. Thus, factory complexes are being designed in clusters—based on the industrial ecology concept—to maximize the circular flow of materials and recycling of wastes among plants. Finally, inter- and intra-generational equity (especially poverty alleviation), pluralistic and consultative decision-making, and enhanced social values and institutions, are important additional aspects that should be considered (at least in the form of safe limits or constraints).

Greenhouse gas mitigation provides an interesting example of how such an integrative framework could help to incorporate climate change response measures within a national sustainable development strategy. The rate of total GHG emissions (G) may be decomposed by means of the following identity:

$$G = (Q/P) \times (Y/Q) \times (G/Y) \times P$$

where (Q/P) is quality of life per capita; (Y/Q) is the material consumption required per unit of quality of life; (G/Y) is the GHG emission per unit of consumption; and P is the population. A high quality of life can be consistent with low total GHG emissions, provided that each of the other three terms on the right-hand side of the identity could be minimized (see also the discussion below on 'tunnelling' and 'leapfrogging'). Reducing (Y/Q) implies 'social decoupling' (or 'dematerialization') whereby satisfaction becomes less dependent on material consumption—through changes in tastes, behaviour and social values. Similarly (G/Y) may be reduced by 'technological decoupling' (or 'decarbonization') that reduces the intensity of GHG emissions in consumption and production. Finally, population growth needs to be reduced, especially where emissions per capita are already high. The linkages between social and technological decoupling need to be explored (see for example, IPCC, 1999). For example, changes in public perceptions and tastes could affect the directions of technological progress, and influence the effectiveness of mitigation and adaptation policies.

Climate change researchers are currently exploring the application of large and complex integrated assessment models or IAMs, which contain coupled submodels that represent a variety of ecological, geophysical and socioeconomic systems (IPCC, 1997). There is considerable scope to examine how both the optimality and durability approaches might be applied in a consistent manner to the various submodels within an IAM, where appropriate.

2.2.10 Cost-benefit Analysis (CBA)

Cost-benefit analysis (CBA) is one well-known example of a single-valued approach, which seeks to assign economic values to the various consequences of an economic activity. The resulting costs and benefits are combined into a single decision-making criterion, such as the net present value (NPV), internal rate of return (IRR), or benefit-cost ratio (BCR). The basic criterion for accepting a project is that the net present value (NPV) of benefits is positive. Typically, NPV = PVB – PVC, where

$$PVB = \sum_{t=0}^{T} B_t/(1+r)^t$$

$$PVC = \sum_{t=0}^{T} C_t/(1+r)^t$$

B_t and C_t are the project benefits and costs in year t, r is the discount rate, and T is the time horizon. Both benefits and costs are defined as the difference between what would occur *with and without* the project being implemented.

When two projects are compared, the one with the higher NPV is deemed superior. Furthermore, if both projects yield the same benefits (PVB), then it is possible to derive the least cost criterion—where the project with the lower PVC is preferred. The IRR is defined as that value of the discount rate for which PVB = PVC, whereas BCR = PVB/PVC. Further details of these criteria, as well as their relative merits in the context of sustainable development, are provided in Munasinghe, 1993.

If a purely financial analysis is required from the private entrepreneurs viewpoint, then B, C, and r are defined in terms of market or financial prices, and NPV yields the discounted monetary profit. This situation corresponds to the economist's ideal world of perfect competition, where numerous profit-maximizing producers and utility-maximizing consumers achieve a Pareto-optimal outcome. However, conditions in the real world are far from perfect, owing to monopoly practices, externalities (such as environmental impacts which are not internalized in the private market), and interference in the market process (*e.g.* taxes). Such distortions cause market (or financial) prices for goods and services to diverge from their economically efficient values. Therefore, the economic efficiency viewpoint usually requires that shadow prices (or opportunity costs) be used to measure B, C and r. In simple terms, the shadow price of a given scarce economic resource is given by the change in value of economic output caused by a unit change in the availability of that resource. In practice, there are many techniques for measuring shadow prices—*e.g.* removing taxes, duties and subsidies from market prices (for details, see Munasinghe, 1993; Squire and van der Tak, 1975).

	Conventional market	Implicit market	Constructed market
Actual behaviour	Effect on production	Travel cost	Artificial market
	Effect on health	Wage differences	
	Defensive or preventive costs	Property values	
		Proxy marketed goods	
Intended behaviour	Replacement cost		Contingent valuation
	Shadow project		

The incorporation of environmental considerations into the economist's single-valued CBA criterion requires further adjustments. All significant environmental impacts and externalities need to be valued as economic benefits and costs. As explained earlier in the section on indicators, environmental assets may be quantified in physical or biological units. Recent techniques for economically valuing environmental impacts are summarized in Box 2.3. However, many of them (such as biodiversity) cannot be accurately valued in monetary terms, despite the progress that has been made in recent years (Munasinghe, 1993; Freeman, 1993). Therefore, criteria like NPV often fail to adequately represent the environmental aspect of sustainable development.

Box 2.3: Recent Techniques for Economically Valuing
Environmental Impacts
(*Source:* Munasinghe, 1993).

Effect on production. An investment decision often has environmental impacts, which in turn affect the quantity, quality or production costs of a range of productive outputs that may be valued readily in economic terms.

Effect on health. This approach is based on health impacts caused by pollution and environmental degradation. One practical measure related to the effect on production is the value of human output lost due to ill health or premature death. The loss of potential net earnings (called the human capital technique) is one proxy for foregone output, to which the costs of health care or prevention may be added.

Defensive or preventive costs. Often, costs may be incurred to mitigate the damage caused by an adverse environmental impact. For example,

---

(contd.)

Box 2.3 (contd.)

if the drinking water is polluted, extra purification may be needed. Then, such additional defensive or preventive expenditures (ex-post) could be taken as a minimum estimate of the benefits of mitigation.

Replacement cost and shadow project. If an environmental resource that has been impaired is likely to be replaced in the future by another asset that provides equivalent services, then the costs of replacement may be used as a proxy for the environmental damage-assuming that the benefits from the original resource are at least as valuable as the replacement expenses. A shadow project is usually designed specifically to offset the environmental damage caused by another project. For example, if the original project was a dam that inundated some forest land, then the shadow project might involve the replanting of an equivalent area of forest, elsewhere.

Travel cost. This method seeks to determine the demand for a recreational site (*e.g.* number of visits per year to a park), as a function of variables like price, visitor income, and socio-economic characteristics. The price is usually the sum of entry fees to the site, costs of travel, and opportunity cost of time spent. The consumer surplus associated with the demand curve provides an estimate of the value of the recreational site in question.

Property Value. In areas where relatively competitive markets exist for land, it is possible to decompose real estate prices into components attributable to different characteristics like house and lot size, air and water quality. The marginal willingness to pay (WTP) for improved local environmental quality is reflected in the increased price of housing in cleaner neighbourhoods. This method has limited application in developing countries, since it requires a competitive housing market, as well as sophisticated data and tools of statistical analysis.

Wage differences. As in the case of property values, the wage differential method attempts to relate changes in the wage rate to environmental conditions, after accounting for the effects of all factors other than environment (*e.g.* age, skill level, job responsibility, etc.) that might influence wages.

Proxy marketed goods. This method is useful when an environmental good or service has no readily determined market value, but a close substitute exists which does have a competitively determined price. In such a case, the market price of the substitute may be used as a proxy for the value of the environmental resource.

Artificial market. Such markets are constructed for experimental purposes, to determine consumer WTP for a good or service. For example, a

(contd.)

Box 2.3 (contd.)

> home water purification kit might be marketed at various price levels, or access to a game reserve may be offered on the basis of different admission fees, thereby facilitating the estimation of values.
>
> **Contingent valuation.** This method puts direct questions to individuals to determine how much they might be willing to pay for an environmental resource, or how much compensation they would be willing to accept if they were deprived of the same resource. The contingent valuation method (CVM) is more effective when the respondents are familiar with the environmental good or service (*e.g.* water quality) and have adequate information on which to base their preferences. Recent studies indicate that CVM, cautiously and rigorously applied, could provide rough estimates of value that would be helpful in economic decision-making, especially when other valuation methods were unavailable.

Capturing the social dimension of sustainable development within CBA is even more problematic. Some attempts have been made to attach 'social weights' to costs and benefits so that the resultant NPV favours poorer groups (see also Box 2.2). However, such adjustments (or preferential treatment for the poor) are rather arbitrary, and have weak foundations in economic theory. Other key social considerations, such as empowerment and participation, are hardly represented within CBA. In summary, the conventional CBA methodology would tend to favour the market-based economic viewpoint, although environmental and social considerations might be introduced in the form of side constraints.

2.2.11 Multi-criteria Analysis (MCA)

Multi-criteria analysis (MCA) or multi-objective decision-making is particularly useful in situations when a single criterion approach like CBA falls short. In MCA, desirable objectives are specified, usually within a hierarchical structure. The highest level represents the broad overall objectives (for example, improving the quality of life), which are often vaguely stated. However, they can be broken down usually into more operationally relevant and easily measurable lower level objectives (*e.g.* increased income). Sometimes only proxies are available—*e.g.* if the objective is to preserve biological diversity in a rainforest, the practically available attribute may be the number of hectares of rainforest remaining. Although value judgements may be required in choosing the proper attribute (especially if proxies are used), actual measurement does not have to be in monetary terms—unlike CBA. More explicit recognition is given to the fact that a variety of objectives and indicators may influence planning decisions.

Figure 2.3 is a two-dimensional representation of the basic concepts underlying MCA. Consider an electricity supplier, who is evaluating a hydroelectric project that could potentially cause biodiversity loss. Objective Z_1 is the additional project cost required to protect biodiversity, and Z_2 is an index indicating the loss of biodiversity. The points A, B, C and D in the Figure represent alternative projects (*e.g.* different designs for the dam). In this case, project B is superior to (or dominates) A in terms of both Z_1 and Z_2—because B exhibits lower costs as well as less bio-diversity loss relative to A. Thus, alternative A may be discarded. However, when we compare B and C, the choice is more complicated since the former is better than the latter with respect to costs but worse with respect to biodiversity loss. Proceeding in this fashion, a trade-off curve (or *locus* of best options) may be defined by all the non-dominated feasible project alternatives such as B, C and D. Such a curve implicitly places both economic and environmental attributes on a more equal footing, in the spirit of sustainomics.

Further ranking of alternatives is not possible without the introduction of value judgements (for an unconstrained problem). Typically, additional information may be provided by a family of equi-preference curves that indicate the way in which the decision-maker or society trades off one objective against the other (see Figure 2.3). Each such equi-preference curve indicates the *locus* of points along which society is indifferent to the trade-off between the two objectives. The preferred alternative is the one that yields the greatest utility—*i.e.* at the point of tangency D of the trade-off curve with the best equi-preference curve (*i.e.* the one closest to the origin).

Because equi-preference curves are usually not measurable, other practical techniques may be used to narrow down the set of feasible choices on the trade-off curve. One approach uses limits on objectives or 'exclusionary screening'. For example, the decision-maker may face an upper bound on costs (*i.e.* a budgetary constraint), depicted by C_{max} in Figure 2.3. Similarly, ecological experts might set a maximum value of biodiversity loss B_{max} (*e.g.* a level beyond which the ecosystem suffers catastrophic collapse). These two constraints may be interpreted in the context of durability considerations, mentioned earlier. Thus, exceeding C_{max} is likely to threaten the viability of the electricity supplier, with ensuing social and economic consequences (*e.g.* jobs, incomes, returns to investors etc.). Similarly, violating the biodiversity constraint will undermine the resilience and sustainability of the forest ecolosystem.

In a more practical sense, C_{max} and B_{max} help to define a more restricted portion of the trade-off curve (darker line)—thereby narrowing and simplifying the choices available to the single alternative D, in Figure 2.3. This type of analysis may be expanded to include other dimensions and attributes. For example, in our hydroelectric dam case, the number of people displaced (or resettled) could be represented by another social variable Z_3.

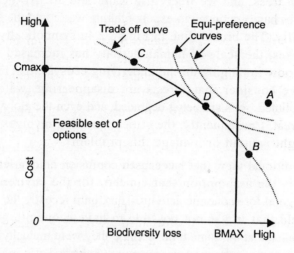

Fig. 2.3: Simple Two-Dimensional Example of Multi-criteria Analysis.

2.2.12 Restructuring Development and Growth for Greater Sustainability

Growth is a major objective of almost all developing countries—especially the poorest ones. This promise cannot be fulfilled unless economic growth is sustained into the long term. The developing countries need to ensure that their endowments of natural resources are not taken for granted and squandered. If valuable resources such as air, forests, soil, and water are not protected, development is unlikely to be sustainable—not just for a few years, but for many decades. Furthermore, on the social side, it is imperative to reduce poverty, create employment, improve human skills and strengthen our institutions.

Next, let us examine the alternative growth paths available, and the role of sustainomics principles in choosing options. Lovelock (1975) made a pioneering contribution with his Gaia hypothesis. He proposed that the totality of life on Earth might be considered an integrated web which works to create a favourable environment for survival. As a corollary, unregulated expansion of human activity might threaten the natural balance. In this spirit, Figure 1.4a shows how the socioeconomic subsystem (solid rectangle) has always been embedded in a broader ecological system (large oval). National economies are inextricably linked to, and dependent on natural resources—since everyday goods and services are in fact derived from natural resources inputs that originate from the larger ecological system. We extract oil from the ground and timber from trees, and we freely use water and air. At the same time, such activities have continued to expel polluting waste into the environment, quite liberally. The broken line in Figure 2.4a symbolically shows that in many cases, the scale of human activity has increased to the point where it is now impinging on the underlying ecosystem. This is evident today, if we consider that forests are disappearing, water resources are being polluted, soils are being degraded, and even the global atmosphere is under threat. Consequently, the critical question involves how human society might contain or manage this problem?

One traditional view that has caused confusion among leaders around the world is the assumption that concern for the environment is not necessarily good for economic activity. Thus, until recently the conventional wisdom held that it was not possible to have economic growth and a good environment at the same time, because they were mutually incompatible goals. However, the more modern viewpoint (embodied also in sustainomics), indicates that growth and environment are indeed complements. One key underlying assumption is that it is often possible to devise so-called 'win-win' policies, which lead to economic as well as environmental gains (Munasinghe *et al.*, 2001). As illustrated earlier in Figure 2.4a, the traditional approach to development would certainly lead to a situation where the economic system would impinge upon the boundaries of the

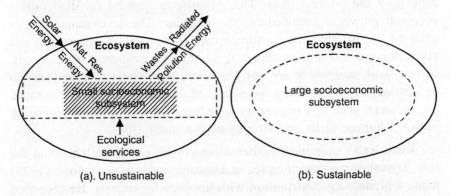

(a). Unsustainable   (b). Sustainable

Fig. 2.4: Restructuring Development to Make the Embedded Socioeconomic Subsystem more Sustainable within the Larger Ecosystem.

ecosystem in a harmful manner. On the other hand, Figure 2.4b summarizes the modern approach that would allow us to have the same level of prosperity without severely damaging the environment. In this case, the oval outer curve is matched by an oval inner curve—where economic activities have been restructured in a way that is more harmonious with the ecosystem.

It would be fruitful to seek specific interventions that might help to make the crucial change in mindset, where the emphasis would be on the structure of development, rather than the magnitude of growth (conventionally measured). Policies that promote environmentally- and socially-friendly technologies that use natural resource inputs more frugally and efficiently, reduce polluting emissions, and facilitate public participation in decision-making, are important. One example is the information technology (IT) revolution, which might facilitate desirable restructuring from an environmental perspective, by making modern economies more services oriented, and shifting activities away from highly polluting and material intensive types of manufacturing and extractive industries (Munasinghe, 1994, 1989). If properly managed, IT could also make development more socially sustainable, by improving access to information, increasing public participation in decision-making, and empowering disadvantaged groups. The correct blend of market forces and regulatory safeguards are required.

2.2.13 Linking Sustainable Development Issues with Conventional Decision-making

Sustainomics helps in identifying practical economic, social and natural resource management options that facilitate sustainable development. It serves as an essential bridge between the traditional techniques of (economic) decision-making and modern environmental and social analysis. In this context, sustainable development assessment (SDA) is an important tool to ensure balanced analysis of both development and sustainability concerns. The 'economic' component of SDA is based on conventional economic and financial analysis (including cost benefit analysis), as described earlier. The other two key components are environmental and social assessment (EA and SA)—see for example World Bank (1998). Poverty assessment is often interwoven with SDA. Economic, environmental and social analyses need to be integrated and harmonized within SDA. Since traditional decision making relies heavily on economics, a first step towards such an integration would be the systematic incorporation of environmental and social concerns into the policy framework of human society.

Figure 2.5 provides an example of how environmental assessment is combined with economic analysis. The right-hand side of the diagram indicates the hierarchical nature of conventional decision-making in a modern society. The global and transnational level consists of sovereign nation states. In the next level are individual countries, each having a multisectored macroeconomy. Various economic sectors (like industry and agriculture) exist in each country. Finally, each sector consists of different subsectors and projects. The usual decision making process on the right side of Figure 2.5 relies on technoengineering, financial and economic analyses of projects and policies. In particular, conventional

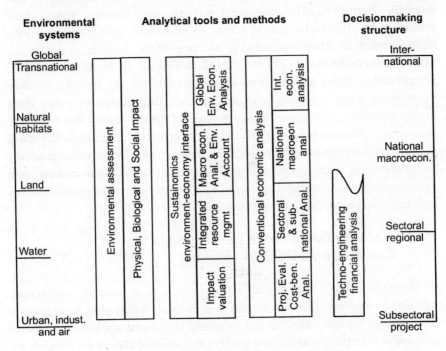

Fig. 2.5: Incorporating Environmental Concerns into Decisionmaking.

economic analysis has been well developed in the past, and uses techniques such as project evaluation/cost-benefit analysis (CBA), sectoral/regional studies, multisectoral macroeconomic analysis, and international economic analysis (finance, trade, etc.) at the various hierarchic levels. Unfortunately, environmental and social analysis cannot be carried out readily using the above decision-making structure. We examine how environmental issues might be incorporated into this framework (with the understanding that similar arguments may be made with regard to social issues). The left side of Figure 2.5 shows one convenient environmental breakdown in which the issues are:

- global and transnational (*e.g.* climate change, ozone layer depletion);
- natural habitats (*e.g.* forests and other ecosystems);
- land (*e.g.* agricultural zone);
- water resource (*e.g.* river basin, aquifer, watershed);
- urban-industrial (*e.g.* metropolitan area, airshed).

In each case, a holistic environmental analysis would seek to study a physical or ecological system in its entirety. Complications arise when such natural systems cut across the structure of human society. For example, a large and complex forest ecosystem (like the Amazon) could span several countries, and also interact with many economic sectors within each country.

The causes of environmental degradation arise from human activity (ignoring natural disasters and other events of non-human origin), and therefore, we begin on the right side of the Figure. The ecological effects of economic decisions must then be traced through to the left side. The techniques of environmental assessment (EA) have been developed to facilitate this difficult analysis (World Bank, 1998). For example, destruction of a primary moist tropical forest may be caused by hydroelectric dams (energy sector policy), roads (transport sector policy), slash and burn farming (agriculture sector policy), mining of minerals (industrial sector policy), land clearing encouraged by land-tax incentives (fiscal policy), and so on. Disentangling and prioritizing these multiple causes (right side) and their impacts (left side) will involve a complex analysis.

Figure 2.5 also shows how sustainomics could play its bridging role at the ecologyeconomy interface, by mapping the EA results (measured in physical or ecological units) onto the framework of conventional economic analysis. A variety of environmental economic techniques including valuation of environmental impacts (at the local/project level), integrated resource management (at the sector/regional level), environmental macroeconomic analysis and environmental accounting (at the economy level), and global/transnational environmental economic analysis (at the international level), facilitate this process of incorporating environmental issues into traditional decision making. Since there is considerable overlap among the analytical techniques described above, this conceptual categorization should not be interpreted too rigidly. Furthermore, when economic valuation of environmental impacts is difficult, techniques such as multicriteria analysis (MCA) would be useful (see Figure 2.3 and earlier discussion on MCA).

Once the foregoing steps are completed, projects and policies must be redesigned to reduce their environmental impacts and shift the development process towards a more sustainable path. Clearly, the formulation and implementation of such policies is itself a difficult task. In the deforestation example described earlier, protecting this ecosystem is likely to raise problems of coordinating policies in a large number of disparate and (usually) non-cooperating ministries and line institutions (*i.e.* energy, transport, agriculture, industry, finance, forestry, etc.).

Analogous reasoning may be readily applied to social assessment (SA) at the society-economy interface, in order to incorporate social considerations more effectively into the conventional economic decision making framework. In this case, the left side of Figure 2.5 would include key elements of SA, such as asset distribution, inclusion, cultural considerations, values and institutions. Impacts on human society (*i.e.* beliefs, values, knowledge and activities), and on the biogeophysical environment (*i.e.* both living and non-living resources), are often linked via second and higher order paths, requiring integrated application of SA and EA. This insight reflects current thinking on the coevolution of socio-economic and ecological systems.

In the framework of the figure, the right side represents a variety of institutional mechanisms (ranging from local to global) which would help to implement policies, measures and management practices to achieve a more sustainable outcome. Implementation of sustainable development strategies and good governance would benefit from the transdisciplinary approach advocated in sustainomics. For example, economic theory emphasises the importance of pricing policy to provide incentives that will influence rational consumer behaviour. However, cases of seemingly irrational or perverse behaviour abound, which might be better understood through findings in areas like behavioural and social psychology, and market research. Such work has identified basic principles that help to influence society and modify human actions, including reciprocity (or repaying favours), behaving consistently, following the lead of others, responding to those we like, obeying legitimate authorities, and valuing scarce resources (Cialdini, 2001).

2.3 APPLYING THE SUSTAINOMICS FRAMEWORK

In this section, practical case studies are presented which illustrate the application of sustainomics principles to make development sustainable at the global-transnational, national, sub-national and local-project scales.

2.3.1 Global-transnational Scale: Climate Change

The climate change problem fits readily within the broad conceptual framework of sustainomics, described above. Decision-makers are beginning to show more interest in the assessment of how serious a threat climate change poses to the future basis for improving human welfare (Munasinghe, 2000; Munasinghe and Swart, 2000). For example, increased GHG emissions and other unsustainable practices are likely to undermine the security of nations and communities, through economic, social and environmental impoverishment, as well as inequitable distribution of adverse impacts—with undesirable consequences such as large numbers of 'environmental' refugees (Lonergan, 1993; Ruitenbeek, 1996; Westing, 1992). Some of the potential linkages, and the sustainomics-related principles and concepts that apply in this context, are outlined below.

2.3.1.1 *Economic, Social and Environmental Risks*

First, global warming poses a significant potential threat to the future economic well being of large numbers of human beings. In its simplest form, the economic efficiency viewpoint will seek to maximize the net benefits (or outputs of goods and services) from the use of the global resource represented by the atmosphere. Broadly speaking, this implies that the stock of atmospheric assets, which provide a sink function for GHGs, needs to be maintained at an optimum level. As indicated in the case study below, this target level is defined at the point where the marginal GHG abatement costs are equal to the marginal avoided damages. The underlying principles are based on optimality and the economically efficient use of a scarce resource, *i.e.* the global atmosphere.

Second, climate change could also undermine social welfare and equity in an unprecedented manner. In particular, more attention needs to be paid to the vulnerability of social values and institutions, which are already stressed due to rapid technological changes (Adger, 1999). Especially within developing countries, erosion of social capital is undermining the basic glue that binds communities together—*e.g.* the rules and arrangements that align individual behaviour with collective goals (Banuri *et al.*, 1994). Existing international mechanisms and systems to deal with transnational and global problems are fragile, and unlikely to be able to cope with worsening climate change impacts.

Furthermore, both intra- and inter-generational equity are likely to be worsened (IPCC, 1996a). Existing evidence clearly demonstrates that poorer nations and disadvantaged groups within nations are especially vulnerable to disasters (Clarke and Munasinghe, 1995; Banuri, 1998). Climate change is likely to result in inequities due to the uneven distribution of the costs of damage, as well as of necessary adaptation and mitigation efforts—such differential effects could occur both among and within countries. Although relevant information is unavailable on global-scale phenomena such as climate change, some historical evidence based on large-scale disasters like El Nino provide useful insights.

Two catastrophic famines or holocausts during the late 19th century killed tens of millions in the developing world. Recent research indicates that they were the outcome of negative synergies between adverse global environmental factors (*i.e.*, the El Nino droughts of 1876-78 and 1898-1901), and the inadequate response of socio-economic systems (*i.e.* vulnerability of tropical farming forcibly integrated into world commodity markets). In the 18th century, the quality of life in countries like Brazil, China, and India was at least on a par with European standards. However, colonial dictates and rapid expansion of world trade re-oriented production in developing countries to service distant European markets. By the time the El Nino droughts struck in the 19th century, the domination of commodity and financial markets by Britain forced developing country smallholders to export at ever-deteriorating terms of trade. This process undermined local food security, impoverished large populations, and culminated in holocausts on an unprecedented scale—identified as one major cause of the present state of underdevelopment in the Third World. From a sustainomics perspective, the corollary is clear, based on the precautionary principle (see next section). The future vulnerability of developing country food production systems to a combination of climate change impacts and accelerated globalization of commodity and financial markets, poses significant risks to the survival of billions, especially in the poorest nations.

Inequitable distributions are not only ethically unappealing, but also may be unsustainable in the long run (Burton, 1997). For example, a future scenario that restricts *per capita* carbon emissions in the South to 0.5 tonnes per year while permitting a corresponding level in the North of over three tonnes per year will not facilitate the cooperation of developing countries, and therefore is unlikely to be durable. More generally, inequity could undermine social cohesion and exacerbate conflicts over scarce resources.

Third, the environmental viewpoint draws attention to the fact that increasing anthropogenic emissions and accumulations of GHGs might significantly perturb a critical global subsystem—the atmosphere (UNFCCC, 1993). Environmental sustainability will depend on several factors, including:

- climate change intensity (*e.g.* magnitude and frequency of shocks);
- system vulnerability (*e.g.* extent of impact damage);
- system resilience (*i.e.* ability to recover from impacts).

Changes in the global climate (*e.g.* mean temperature, precipitation, etc.) could also threaten the stability of a range of critical, interlinked physical, ecological and social systems and subsystems (IPCC, 1996b).

2.3.1.2 Relevant Principles for Policy Formulation

When considering climate change response options, several principles and ideas that are widely used in environmental economics analysis would be useful—these include the polluter pays principle, economic valuation, internalization of externalities, and property rights. The polluter pays principle argues that those who are responsible for damaging emissions should pay the corresponding costs. The economic rationale is that this provides an incentive for polluters to reduce their emissions to optimal (*i.e.* economically efficient) levels. Here, the idea of economic valuation becomes crucial. Quantification and economic valuation of potential damage from polluting emissions is an important prerequisite. In the case of a common property resource like the atmosphere, GHG emitters can freely pollute without penalties. Such 'externalities' need to be internalized by imposing costs on polluters that reflect the damage caused. An externality occurs when the welfare of one party is affected by the activity of another party who does not take these repercussions into account in his/her decision-making (*e.g.* no compensating payments are made). The theoretical basis for this is well known since Pigou (1932) originally defined and treated externalities in rigorous fashion. In this context, the notion of property rights is also relevant to establish that the atmosphere is a valuable and scarce resource that cannot be used freely and indiscriminately.

An important social principle is that climate change should not be allowed to worsen existing inequities—although climate change policy cannot be expected to address all prevailing equity issues. Some special aspects include:

- the establishment of an equitable and participative global framework for making and implementing collective decisions about climate change;
- reducing the potential for social disruption and conflicts arising from climate change impacts;
- protection of threatened cultures and preservation of cultural diversity.

From the social equity viewpoint, the polluter pays principle is based not only on economic efficiency, but also on fairness. An extension of this idea is the principle of recompensing victims—ideally by using the revenues collected from polluters. There is also the moral/equity issue concerning the extent of the polluters' obligation to compensate for past emissions (*i.e.* a form of environmental debt). As mentioned earlier, weighing the benefits and costs of climate change impacts according to the income levels of those who are affected, has also been suggested as one way of redressing inequitable outcomes. Kverndokk (1995) argued that conventional justice principles would favour the equitable allocation of future GHG emission rights on the basis of population. Equal *per capita* GHG emission rights (*i.e.* equal access to the global atmosphere) is consistent also with the UN human rights declaration underlining the equality of all human beings.

Traditionally, economic analysis has addressed efficiency and distributional issues separately—*i.e.* the maximization of net benefits is distinct from who might receive such gains. Recent work has sought to interlink efficiency and equity more naturally. For example, environmental services could be considered public goods, and incorporated into appropriate markets as privately produced public goods (Chichilnisky and Heal, 2000). Some social equity and economic efficiency interactions are discussed in Box 2.2.

Several concepts from contemporary environmental and social analysis are relevant for developing climate change response options, including the concepts of durability, optimality, safe limits, carrying capacity, irreversibility, non-linear responses, and the precautionary principle. Broadly speaking, durability and optimality are complementary and potentially convergent approaches (see earlier discussion). Under the durability criterion, an important goal would be to determine the safe limits for climate change within which the resilience of global ecological and social systems would not be seriously threatened. In turn, the accumulations of GHGs in the atmosphere would have to be constrained to a point that prevented climate change from exceeding these safe margins. It is considered important to avoid irreversible damage to bio-geophysical systems and prevent major disruption of socioeconomic systems. Some systems may respond to climate change in a non-linear fashion, with the potential for catastrophic collapse. Thus, the precautionary principle argues that lack of scientific certainty about climate change effects should not become a basis for inaction, especially where relatively low cost steps to mitigate climate change could be undertaken as a form of insurance (UNFCCC, 1993).

2.3.1.3 Case Study 1: The Interplay of Optimality and Durability in Determining Appropriate Global GHG Emission Target Levels

Optimization and durability based approaches can facilitate the determination of target GHG emission levels (Munasinghe, 1998a). Under an economic optimizing framework, the ideal solution would be first to estimate the long-run marginal abatement costs (MAC) and the marginal avoided damages (MAD) associated with different GHG emission profiles—see Figure 2.6c, where the error bars on the curves indicate measurement uncertainties (IPCC, 1996a). The optimal emission levels would be determined at the point where future benefits (in terms of climate change damage avoided by reducing one unit of GHG emissions) are just equal to the corresponding costs (of mitigation measures required to reduce that unit of GHG emissions), *i.e.* MAC = MAD at point R_{OP}.

Durable strategies become more relevant when we recognize that MAC and/or MAD might be poorly quantified and uncertain. Figure

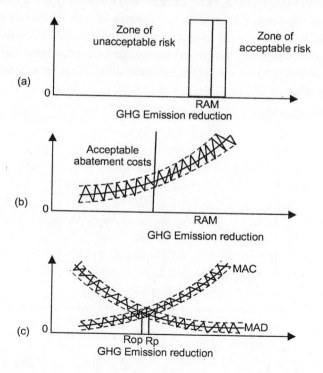

Fig. 2.6: Determining Abatement Targets: (a) Absolute Standard; (b) Affordable Safe Minimum Standard; (c) Cost-benefit Optimum
Source: Adapted from IPCC 1996c, Figure 5.10.

2.6b assumes that MAC is better defined than MAD. First, MAC is determined using techno-economic least-cost analysis—an optimizing approach. Next, the target emissions are set on the basis of the affordable safe minimum standard (at R_{AM}), which is the upper limit on costs that will still avoid unacceptable socioeconomic disruption—this is closer to the durability approach.

Finally, Figure 2.6a indicates an even more uncertain world, where neither MAC nor MAD is defined. Here, the emission target is established on the basis of an absolute standard (R_{AS}) or safe limit, which would avoid an unacceptably high risk of damage to ecological (and/or social) systems. This last approach would be more in line with the durability concept.

2.3.1.4 Case Study 2: Combining Efficiency and Equity to Facilitate South-North Cooperation for Climate Change Mitigation

GHG mitigation efforts will require worldwide cooperation. Figure 2.7 clarifies the basic rationale for greater North to South resource transfers and technical cooperation, and also highlights how the sustainomics approach elucidates the complex interaction of economic efficiency, social equity and global environmental considerations in addressing the climate change problem. The curve ABCDE indicates the combined marginal abatement costs (MAC) for a pair of countries (one developing

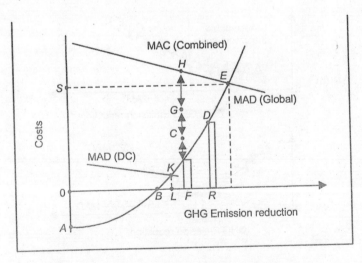

Fig. 2.7: Rationale for South-North Cooperation
and Interplay of Efficiency and Equity.
Source: Munasinghe and Munasinghe (1993)

or southern and the other industrialized or northern). In other words, the graph shows the additional costs of adopting various GHG-reducing schemes (over and above the costs of conventional technologies), plotted against the amount of avoided emissions. The portion AB indicates negative costs, to represent so-called 'win-win' or 'no regrets' options—like energy efficiency schemes for which cost-benefit analysis will show a net economic gain even before GHG abatement benefits have been considered (*i.e.* where the value of conventional energy savings exceed project costs).

Other measures like fuel switching, new and renewable technologies, carbon sinks, and advanced energy technologies are likely to appear on the rising part (BCDE) of the curve. Many lower cost options for GHG emissions reduction (such as CF), would be in the developing country, whereas more costly alternatives would lie in the industrialized nation.

On its own, a typical developing country would be willing to pursue abatement measures only up to the point *K*—where MAC is equal to the benefit of avoided climate change costs or MAD(DC) accruing purely to that country. Ideally, all options should be pursued in both countries, up to the point *E*, where the additional costs (MAC combined) of the marginal unit of emissions curtailed are equal to the corresponding benefits (MAD global) of avoided global warming impacts. Although the benefits curves will not be known with precision, the precautionary principle and the high risk involved would suggest that the point *E* would be far to the right of *K*.

First, we explore the implications of this broad environmental rationale for resource transfers from the North to the South. In this context, consider a representative GHG mitigation project (*e.g.* re-afforestation) in the developing country, where the additional costs of GHG emissions reduction is CF. It would be *economically efficient* for the global community to finance these costs (on a grant basis) in the developing country, because they will thereby realize the global net benefits HC (*i.e.* HC = HF − CF). This would effectively internalize the global environmental externality.

Second, we make the case for a bilateral transfer of resources from an industrialized to a developing country. Consider the cost of a project DR (*e.g.* conversion of coal plants), which seeks to reduce GHG emissions in the industrialized country. This country could realize a cost saving GC by transferring an amount CF to the developing country, while still achieving the same global emissions reduction. The foregoing could be the basis for bilateral cooperative schemes such as joint implementation (JI) and/or the clean development mechanism (CDM), under the Kyoto Protocol. To the extent that net benefits HC, and cost savings GC are significant, it would be both *equitable* and *efficient* for the industrial nation to give the poorer developing country more resources than the (minimum) breakeven reimbursement CF. In other words, the equity principle of sustainomics would favour the sharing of cost savings GC between the two cooperating nations. The underlying ethical argument would be based on the facts that:

- both the historical and current levels of *per capita* GHG emissions from the industrial country are likely to be many times the corresponding contribution from the developing nation;
- the *per capita* income and ability to pay of the industrial country would be many times greater than those of the developing country.

This would also provide a greater incentive for the developing country to participate in such a scheme. The same argument has been made in the case of South-North cooperation to reduce ozone-depleting substances under the Montreal Protocol (Munasinghe and King, 1992).

1.3.2 National-economy Scale: Macroeconomic Management

Conventional economic valuation of environmental impacts is a key step in incorporating the results of project level environmental assessment into economic decision-making—*e.g.* cost-benefit analysis (see also Figure 2.5 and associated discussion). At the macroeconomic level, recent work has focussed on incorporating environmental considerations such as depletion of natural resources and pollution damage into the system of national accounts (UN Statistical Office, 1993; Atkinson *et al.*, 1997). These efforts have yielded useful new indicators and measures such as the system of environmentally adjusted environmental accounts (SEEA), green gross national product, and genuine savings, which adjust conventional macroeconomic measures to allow for environmental effects.

Meanwhile, national policy-makers routinely make many key macro-level decisions that could have (often inadvertent) environmental and social impacts, which are far more significant than the effects of local economic activities. These pervasive and powerful measures are aimed at achieving economic development goals like accelerated growth—which invariably have a high priority in national agendas. Typically, many macroeconomic policies seek to induce rapid growth, which in turn could potentially result in greater environmental harm or impoverishment of already disadvantaged groups. More attention needs to be paid to such economy policies, whose environmental and social linkages have not been adequately explored in the past (Munasinghe and Cruz, 1994).

Clearly, sustainable development strategies that are consistent with other national development policies are more likely to be effective than isolated technological or policy options. In particular, the highest priority needs to be given to finding 'win-win policies', which not only achieve conventional macroeconomic objectives, but also make local and national development efforts more sustainable. Such policies could help to build support for sustainable development strategies among the traditional decision-making community, and conversely make sustainable development specialists more sensitive to shorter term macroeconomic needs. They would reduce the potential for conflict between two powerful current trends—the growth oriented, market based economic reform process, and protection of the environment.

2.3.2.1 Scope of Policies and Range of Impacts

The most powerful economic management tools currently in common use are economy reforms, which include structural adjustment packages. Economy (or countrywide) policies consist of both sectoral and macroeconomic policies which have widespread effects throughout the economy. Sectoral measures mainly involve a variety of economic instruments, including pricing in key sectors (for example, energy or agriculture) and broad sectorwide taxation or subsidy programmes (for example, agricultural production subsidies, and industrial investment incentives). Macroeconomic measures are even more sweeping, ranging from exchange rate, interest rate, and wage policies, to trade liberalization, privatization, and similar programs. Since space limitations preclude a comprehensive review of interactions between economy policies and sustainable development, we briefly examine several examples that provide a flavour of the possibilities involved (for details, see Munasinghe, 1997; Jepma and Munasinghe, 1998).

On the positive side, liberalizing policies such as the removal of price distortions and promotion of market incentives have the potential to improve economic growth rates, while increasing the value of output per unit of pollution emitted (*i.e.* so called 'win-win' outcomes). For example, reforms that improve the efficiency of energy use could reduce economic waste and lower the severity of air pollution. Similarly, improving property rights and strengthening incentives for better land management not only yield economic gains but also reduce deforestation of open access lands (*e.g.* due to slash and burn agriculture).

At the same time, growth-inducing economic policies could lead to increased environmental and social damage, unless the macro-reforms are complemented by additional environmental and social measures. Such negative impacts are invariably unintended and occur when some broad policy changes are undertaken while other hidden or neglected economic and institutional imperfections persist (Munasinghe and Cruz, 1994). In general, the remedy does not require reversal of the original reforms, but rather the implementation of additional complementary measures (both economic and noneconomic) that mitigate climate change. For example, export promotion measures and currency devaluation might increase the profitability of timber exports (see the case study below). This in turn could further accelerate deforestation that was already under way due to low stumpage fees and open access to forest lands. Establishing property rights and increasing timber charges would reduce deforestation, without interrupting the macroeconomic benefits of trade liberalization.

Similarly, market-oriented liberalization could lead to economic expansion and the growth of wasteful energy-intensive activities in a country where subsidized energy prices persisted. Eliminating the energy price subsidies could help to reduce local air pollution and net GHG emissions while enhancing macroeconomic gains. Countrywide policies could also influence adaptation to climate change, negatively or positively. For example, national policies that encouraged population movement into low-lying coastal areas might increase their vulnerability to future impacts of sea-level rise. On the other hand, government actions to protect citizens from natural disasters—such as investing in safer physical infrastructure or strengthening the social resilience of poorer communities—could help to reduce vulnerability to extreme weather events associated with future climate change (Clarke and Munasinghe, 1995).

In this context, the sustainomics approach helps to identify and analyse economic-environmental-social interactions, and formulate effective sustainable development policies, by linking and articulating these activities explicitly. Implementation of such an approach would be facilitated by constructing a simple Action Impact Matrix or AIM, as described below in Case Study 3 (Munasinghe and Cruz, 1994).

2.3.2.2 Case Study 3: Action Impact Matrix (AIM) for Policy Analysis

The sustainomics approach seeks to identify and analyse economic-environmental-social interactions, and thereby formulate more sustainable development policies. One tool that would facilitate the implementation of such an approach is the Action Impact Matrix (AIM)—a simple example is shown in Table 2.1, although an actual AIM would be very much larger and more detailed (Munasinghe, 1993, 1998b). Such a matrix helps to promote an integrated view, meshing development decisions with priority economic, environmental and social impacts. The far left column of the table lists examples of the main development interventions (both policies and projects), while the top row indicates some typical sustainable development issues. Thus the elements or cells in the matrix help to:

- identify explicitly the key linkages;
- focus attention on methods of analysing the most important impacts;
- suggest action priorities and remedies.

At the same time, the organization of the overall matrix facilitates the tracing of impacts, as well as the coherent articulation of the links among a range of development actions—both policies and projects.

A stepwise procedure, starting with readily available data, has been used effectively to develop the AIM in several country studies (Munasinghe and Cruz, 1994). This process has helped to harmonize views among those involved (economists, ecologists, sociologists and others), thereby improving the prospects for successful implementation.

Screening and Problem Identification: One of the early objectives of the AIM-based process is to help in *screening and problem identification*— by preparing a preliminary matrix that identifies broad relationships, and provides a qualitative idea of the magnitudes of the impacts. Thus, the preliminary AIM would be used to prioritize the most important links between policies and their sustainability impacts. For example, in row 2 of Table 2.1, a currency devaluation aimed at improving the trade balance may make timber exports more profitable and lead to deforestation of open access forests. Column 3 indicates severe land degradation and biodiversity. Lower down in the same column, one appropriate remedy might involve complementary measures to strengthen property rights and restrict access to forest areas.

A second example shown in row 3 involves increasing energy prices closer to marginal costs—to improve energy efficiency, while decreasing air pollution and GHG emissions. A complementary measure indicated in column 4 consists of adding pollution taxes to marginal energy costs, which will further reduce air pollution and GHG emissions. Increasing public sector accountability will reinforce favourable responses to these price incentives, by reducing the ability of inefficient firms to

Table 2.1: A Simplified Preliminary Action Impact Matrix (AIM)
(*Source*: Munasinghe and Cruz, 1994).

Activity/policy[a]	Main objective	Impacts on key sustainable development issues			
		Land degradation, biodiversity loss	Air pollution GHG emissions effects	Resettle-ment and social	other
Macroeconomic and sectoral policies	Macroeconomic and sectoral improvements	Positive impacts due to removal of distortions Negative impacts mainly due to remaining constraints			
Exchange rate	Improve trade balance and economic growth	(-H) (deforest open-access areas)			
Energy pricing	Improve energy use economic and efficiency		(+M) (energy efficiency)		
Others					
Complementary measures and remedies[b]	Specific socioeconomic and environmental gains	Enhance positive impacts and mitigate negative impacts (above) of broader macroeconomic and sectoral policies			
Market based			(+M) (pollution tax)		

(contd.)

Table 2.1 (contd.)

Non-market based	(+H) (property rights)	(+M) (public sector accountability)	
Investment Projects	Improve effectiveness of investments	Investment decisions made more consistent with broader policy and institutional framework	
Project 1 (*Hydro dam*)	(-H) (inundate forests)	(+M) (displace fossil fuel use)	(-M) (displace people)
Project 2 (*Re-forest and relocate*)	(+H) (replant forests)		(+M) (relocate people)
Project N			

[a] A few examples of typical policies and projects as well as key economic, environmental and social issues are shown. Some illustrative but qualitative impact assessments are also indicated: thus + and – signify beneficial and harmful impacts, while H and M indicate high and moderate intensity. The AIM process helps to focus on the highest priority socioeconomic and environmental issues.

[b] Commonly used market-based measures include effluent charges, tradable emission permits, emission taxes or subsidies, bubbles and offsets (emission banking), stumpage fees, royalties, user fees, deposit-refund schemes, performance bonds, and taxes on products (such as fuel taxes). Non-market based measures comprise regulations and laws specifying environmental standard (such as ambient standards, emission standards, and technology standards) which permit or limit certain actions ('dos' and 'don'ts').

pass on cost increases to consumers or to transfer their losses to the government. In the same vein, a major hydroelectric project is shown lower down in the Table as having two adverse impacts (inundation of forested areas and village dwellings), as well as one positive impact (the replacement of thermal power generation, thereby reducing air pollution and GHG emissions). A re-forestation project coupled with resettlement schemes may help address the negative impacts.

This matrix-based approach therefore encourages the systematic articulation and coordination of policies and projects to make development more sustainable. Based on readily available data, it would be possible to develop such an initial matrix for many countries.

Analysis and Remediation: This process may be developed further to assist in *analysis* and *remediation*. For example, more detailed analyses and modelling may be carried out for those matrix elements in the preliminary AIM that had been already identified as representing high priority linkages between economywide policies and economic, environmental and social impacts. This, in turn, would lead to a more refined and updated AIM, which would help to quantify impacts and formulate additional policy measures to enhance positive linkages and mitigate negative ones.

The types of more detailed analysis that could help to determine the final matrix would depend on planning goals and available data and resources. They may range from fairly simple methods to rather sophisticated economic, ecological and social models, in the sustainomics toolkit.

2.3.2.3 Case Study 4: Restructuring Growth to Address Climate Change Issues

Economic growth continues to be a widely pursued objective of most governments, and therefore, the sustainability of long term growth is a key issue (Munasinghe *et al.*, 2001)—in particular, reducing the intensity of GHG emissions of human activities is an important step in mitigating climate change (Munasinghe, 2000). Given that the majority of the world population lives under conditions of absolute poverty, a climate change strategy that unduly constrained growth prospects in those areas would be more unattractive. A sustainomics based approach would seek to identify measures that modify the structure of development and growth rather than restricting it (see Figure 2.5), so that GHG emissions are mitigated and adaptation options enhanced.

The above approach is illustrated in Figure 2.8, which shows how a country's GHG emissions might vary with its level of development. One would expect carbon emissions to rise more rapidly during the early stages of development (along AB), and begin to level off only when *per capita* incomes are higher (along BC). A typical developing country would be at a point such as B on the curve, and an industrialized nation might be at C. The key point is that if the developing countries were to follow the growth path of the industrialized world, then atmospheric concentrations of GHGs would soon rise to dangerous levels. The risk of exceeding the safe limit (shaded area) could be avoided by adopting sustainable development strategies that would permit developing countries to progress along a path such as BD (and eventually DE), while also reducing GHG emissions in industrialized countries along CE.

As outlined earlier, growth-inducing economywide policies could combine with imperfections in the economy to cause environmental harm. Rather than halting economic growth, complementary policies may be used to remove such imperfections and thereby protect the environment. It would be fruitful to encourage a more proactive approach whereby the developing countries could learn from the past experiences of the industrialized world—by adopting sustainable development strategies and climate change measures which would enable them to follow development paths such as BDE, as shown in Figure 2.8 (Munasinghe, 1998b). Thus, the emphasis is on identifying policies that will help delink carbon emissions and growth, with the curve in Figure 2.8 serving mainly as a useful metaphor or organizing framework for policy analysis.

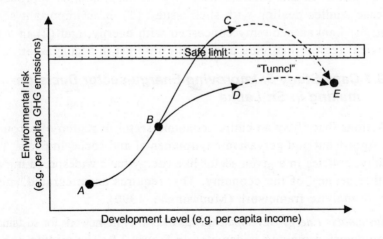

Fig. 2.8: Environmental Risk Versus Development Level
Source: Adapted from Munasinghe [1995].

This representation also illustrates the complementarity of the optimal and durable approaches discussed earlier. It has been shown that the higher path ABC in Figure 2.8 could be caused by economic imperfections which make private decisions deviate from socially optimal ones (Munasinghe, 1998c). Thus the adoption of corrective policies that reduce such divergences from optimality and thereby reduce GHG emissions per unit of output would facilitate movement along the lower path ABD. Concurrently, the durability viewpoint suggests that flattening the peak of environmental damage (at C) would be especially desirable to avoid exceeding the safe limit or threshold representing dangerous accumulations of GHGs (shaded area in Figure 2.8).

Several authors have econometrically estimated the relationship between GHG emissions and *per capita* income using cross-country data and found curves with varying shapes and turning points (Holtz-Eakin and Selden, 1995; Sengupta, 1996; Unruh and Moomaw, 1998; Cole *et al.*, 1997). One reported outcome is an inverted U-shape (called the environmental Kuznets curve or EKC)—like the curve ABCE in the Figure. In this case, the path BDE (both more socially optimal and durable) could be viewed as a sustainable development 'tunnel' through the EKC (Munasinghe, 1995, 1998c).

2.3.3 Sub-national Scale: Energy-sector Planning and Forest Ecosystem Management

At the sub-national scale, sustainable development issues arise in various forms. In this section, we apply the sustainomics approach to two case studies dealing with such issues: (1) in an important sector of the Sri Lankan economy concerned with energy; and (2) in a key ecological region involving a tropical rainforest in Madagascar.

2.3.3.1 Case Study 5: Improving Energy-sector Decision-making in Sri Lanka

Actions that affect an entire economic sector or region of a country have significant and pervasive environmental and social impacts. Thus typically, policies in a given sector like energy have widespread impacts on other sectors of the economy. This requires an integrated, multi-sectoral analytic framework (Munasinghe, 1990).

Sustainable Energy Development Framework: A framework for sustainable energy decision-making is depicted in Figure 2.9. The middle column of the Figure shows the core of the framework comprising an integrated multilevel analysis that can accommodate issues ranging from the global

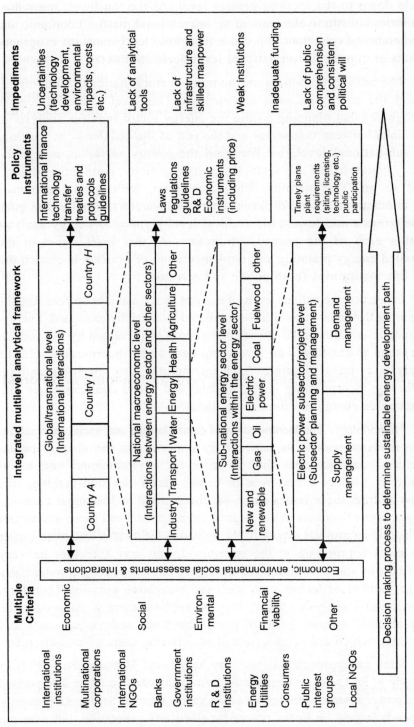

Fig. 2.9: Framework for Sustainable Energy Development
Source: Adapted from Munasinghe [1990].

scale down to the local or project level. At the top level, individual countries constitute elements of an international matrix. Economic and environmental conditions imposed at this global level constitute exogenous inputs or constraints on national level decision-makers.

The next level focuses on the multi-sectoral national economy, of which the energy sector is one element. This level of the framework recognizes that planning within the energy sector requires analysis of the links between that sector and the rest of the economy. At the third or sub-national level, we focus on the energy sector as a separate entity composed of sub-sectors such as electricity, petroleum products and so on. This permits detailed analysis, with special emphasis on interactions among different energy sub-sectors. Finally, the most disaggregate and lowest hierarchical level pertains to energy analysis within each of the energy sub-sectors. At this level, most of the detailed energy planning and implementation of projects is carried out by line institutions (both public and private).

In practice, the various levels of analysis merge and overlap considerably, requiring that inter-sectoral linkages should be carefully examined. Energy-economic-environmental- social interactions (represented by the vertical bar) tend to cut across all levels and need to be incorporated into the analysis as far as possible. Such interactions also provide important paths for incorporating environmental and social considerations into sustainable energy development policies.

Methodology: The incorporation of environmental and social externalities into decision-making is particularly important in the electric power sector. It is also clear that in order for environmental and social concerns to play a real role in power sector decision-making, one must address these issues early—at the sectoral and regional planning stages, rather than later at the stage of environmental and social assessment of individual projects. Many of the valuation techniques discussed earlier are most appropriate at the microlevel, and may therefore be very difficult to apply in situations involving choices among a potentially large number of technology, site, and mitigation options. Therefore, multicriteria analysis (MCA) may be applied, since it allows for the appraisal of options with different objectives and varied costs and benefits, which are often assessed in differing units of measurement.

Such an approach was used by Meier and Munasinghe (1994) in a study of Sri Lanka, to demonstrate how externalities could be incorporated into power system planning in a systematic manner. Sri Lanka presently depends largely on hydro power for electricity generation, but over the next decade the main choices seem to be large coal- or oil-fired stations, or hydro plants whose economic returns and environmental impacts are increasingly unfavourable. In addition, there is a wide range of other options (such as wind power, increasing use of demand side management, and system efficiency improvements), that make decision-making quite difficult—even in the absence of unusual environmental concerns. The study is in its focus on system-wide planning issues, as opposed to the more usual policy of assessing environmental concerns only at the project level after the strategic sectoral development decisions have already been made.

The methodology involves the following steps: (a) definition of the generation options and their analysis using sophisticated least-cost system planning models; (b) selection and definition of the attributes, selected to reflect planning objectives; (c) explicit economic valuation of those impacts for which valuation techniques can be applied with confidence—the resultant values are then added to the system costs to define the overall attribute relating to economic cost; (d) quantification of those attributes for which explicit economic valuation is inappropriate, but for which suitable quantitative impact scales can be defined; (e) translation of attribute value levels into value functions (known as 'scaling'); (f) display of the trade-off space, to facilitate understanding of the trade-offs to be made in decision-making; and (g) definition of a candidate list of options for further study; this also involves the important step of eliminating inferior options from further consideration.

Main Results: The main set of sectoral policy options examined included: (a) variations in the currently available mix of hydro, and thermal (coal and oil) plants; (b) demand-side management (using the illustrative example of compact fluorescent lighting); (c) renewable energy options (using the illustrative technology of wind generation); (d) improvements in system efficiency (using more ambitious targets for transmission and distribution losses than the base case assumption of 12% by 1997); (e) clean coal technology (using pressurized fluidized bed combustion (PFBC) in a combined cycle mode as the illustrative technology); and (f) pollution control technology options (illustrated by a variety of fuel switching and pollution control options such as using imported low sulfur oil for diesels, and fitting coal-burning power plants with flue gas desulfurization (FGD) systems).

Great care needs to be exercised in criteria or attribute selection—they should reflect issues of national as well as local project level significance, and ought to be limited in number. To capture the potential impact on global warming, CO_2 emissions were defined as the appropriate proxy. Health impacts were measured through population-weighted increments in both fine particulates and NO_x attributable to each source. To capture the potential biodiversity impacts, a probabilistic index was derived (see Box 2.4 for details). As an illustrative social impact, employment creation was used.

Figure 1.10a illustrates a typical trade-off curve for biodiversity (see also, the earlier discussion on MCA in the context of Figure 2.3). The 'best' solutions lie closest to the origin. The so-called trade-off curve is defined by the set of 'non-inferior' solutions, representing the set of options that are superior, regardless of the weights assigned to the different objectives. For example, on this curve, the option defined as 'no hydro' is better than the option 'wind', in terms of both economic cost and biodiversity loss.

While most of the options have an index value that falls in the range of 50–100, the no-hydro option has an essentially zero value, because the thermal projects that replace hydro plants in this option tend to lie at sites of poor biodiversity value (either close to load centres or on the coast). Meanwhile, wind plants would require rather large land area, and their biodiversity index is higher. However, the vegetation in the area on the south coast (where the wind power plants would be located) has relatively low biodiversity value, and therefore the overall biodiversity impact of this option is small. In summary, the best options (on the trade-off curve) include the no-hydro, and run-of-river hydro options that require essentially zero inundation. Note the extreme outlier at the top right hand corner, which is the Kukule hydro dam—it has a biodiversity loss index ($B = 530$) that is an order of magnitude larger than for other options ($B = 50$ to 70).

Box 2.4: Developing a Preliminary
Biodiversity Index

In electric power plant evaluation, detailed site-specific information at potential sites is unlikely to be available at the long-range system planning stage. Thus, the only quantification of biodiversity impacts that appears possible at this level of aggregation is a probabilistic estimate that gives the decision-maker advance information about the likelihood that a more detailed environmental impact assessment will reveal adverse effects on an endemic species, significant impacts

---

(contd.)

Box 2.4 (contd.)

on ecosystems of high biological diversity, or degradation of a habitat already in a marginal condition. It should be noted that endemicity and biodiversity are not necessarily correlated, since an endemic species may be encountered in an area of low biodiversity, and areas of high biodiversity may contain no endemic species. However, endemic species in Sri Lanka are most likely to be encountered in areas of high biodiversity.

A biodiversity index must reflect several key characteristics. First is the nature of the impacted system itself. In Table B4, the main agro-ecological zones encountered in Sri Lanka are ranked and assigned a value (*wj*) that captures the relative biodiversity value of different habitats. The scale is to be interpreted as a strict ratio scale (*i.e.* zero indicates zero amount of the characteristic involved, and a habitat value of 0.1 implies ten times the value of a habitat assigned the value of 0.01). The second element concerns the *relative* valuation, because the *value* of the area lost is a function of the proportion of the habitat that is lost. For example, the loss of the *last* hectare of an ecosystem would be unacceptable, and hence assigned a very large value (even if the habitat involved were of low biodiversity, such as a sand dune) whereas the loss of one hectare out of 10,000 ha would be much less valuable.

The total biodiversity index value associated with site i, is defined as:

$$E_i = \sum_j w_j A_{ij}$$

where A_{ij} is the ha of ecosystem of type *j* at site *i*, and w_j is relative biodiversity value of type *j* (as defined in Table B4).

Because E_i would tend to be correlated with reservoir size (*i.e.* land area inundated and energy-storage capacity), two further scaled indices may be defined as follows:

$$F_i = E_i / \left[\sum_j A_{ij} \right] = E_i / \text{[total land area affected at site } i\text{]}$$

$G_i = E_i /$ [hydroelectric energy generated per year at site *i*]

Thus, F_i is the average biodiversity index value per hectare of affected land, and G_i is the average biodiversity index value per unit of energy produced per year.

Relative Biodiversity Values of Agro-ecological Zones
in Sri Lanka (Adapted from Meier and Munasinghe, 1994).

Rank	Ecosystem	Relative biodiversity value
1	Lowland wet evergreen forest	0.98
2	Lowland moist evergreen forest	0.98
3	Lower montane forest	0.90
4	Upper montane forest	0.90
5	Riverline forest	0.75
6	Dry mixed evergreen forest	0.5
7	Villus	0.4
8	Mangroves	0.4
9	Thorn forest	0.3
10	Grasslands	0.3
11	Rubber lands	0.2
12	Home gardens	0.2
13	Salt marshes	0.1
14	Sand dunes	0.1
15	Coconut lands	0.01

A quite different trade-off curve was derived between health impacts and average incremental cost, as illustrated in Figure 2.10b. Note that the point 'iresid' on the trade-off curve (which calls for the use of low sulfur imported fuel oil at diesel plants), is better than the use of flue gas desulfurization systems (point 'FGD')—in terms of both economic cost and environment.

Conclusions: The case study draws several useful conclusions.

First, the results indicate that those impacts for which valuation techniques are relatively straightforward and well-established—such as valuing the opportunity costs of lost production from inundated land, or estimating the benefits of establishing fisheries in a reservoir—tend to be quite small in comparison with overall system costs, and their inclusion into the benefit-cost analysis does not materially change results.

Second, even in cases where explicit valuation may be difficult, such as in the case of mortality and morbidity effects of air pollution, implicit valuation based on analysis of the trade-off curve can provide important guidance to decision-makers.

Third, the case study indicated that certain options were in fact clearly inferior, or clearly superior, to all other options when one examines all impacts simultaneously. For example, the high dam version of the Kukule hydro project can be safely excluded from all further consideration here, as a result of poor performance on all attribute scales (including the economic one). Fourth, the results indicate that it is possible to derive attribute scales that can be useful proxies for impacts that may be difficult to value. For example, use of the population-weighted incremental ambient air pollution scale as a proxy for health impacts permitted a number of important conclusions that are independent of the specific economic value assigned to health effects.

Finally, with respect to the practical implications for planning, the study identified several specific recommendations on priority options, including (i) the need to systematically examine demand side management options, especially fluorescent lighting; (ii) the need to examine whether the present transmission and distribution loss reduction target of 12% ought to be further reduced; (iii) the need to examine the possibilities of pressurized fluidized bed combustion technology for coal power; (iv) replacement of some coal-fired power plants (on the South coast) by diesel units; and (v) the need to re-examine cooling system options for coal plants.

2.3.3.2 Case Study 6: Rainforest Management in Madagascar

Madagascar is one of the economically poorest and ecologically richest countries in the world, and it has been designated by the international community as a prime area for biodiversity whose ecosystems are also at great risk. The government of Madagascar is also taking steps to control forest degradation and to protect biodiversity. The results summarized below are from the first stage in the analysis to arrive at a rational decision concerning the proposed creation of the Mantadia National Park in Madagascar (Kramer *et al.*, 1995).

The establishment of a national park generates many indirect and direct costs and benefits. Costs arise from land acquisition (if the land had been previously privately owned), the hiring of park personnel, and the development of roads, visitors' facilities, and other infrastructure. Another important set of costs that are often ignored are the opportunity costs associated with the foregone uses of park land. Benefits include both use values and non-use values. Tourism can generate considerable revenues for the country from both entrance fees and travel expenditures. National parks also generate a number of non-use benefits, among which existence value and option value are important. Other benefits may include reduced deforestation, watershed protection and climate regulation.

This study measured some of the more important and difficult to measure economic impacts (including the impact of the park on local villagers and the benefits of the new park to foreign tourists), using the techniques summarized earlier in Box 2.3. Local people use the park area for rice cultivation and for gathering forest products. The creation of the park results in an opportunity cost in terms of lost

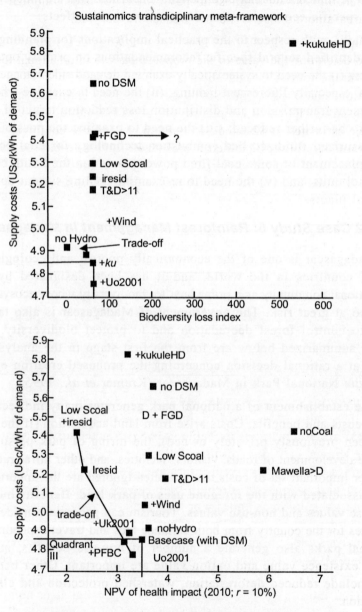

Fig. 2.10: Trade-off Curves Between Economic Costs and
(a) Health Impacts; and (b) Biodiversity Impacts.
Source : Meier and Munasinghe, [1994]

production as presented in Table 2.2—based on detailed surveys of 351 households in 17 villages with a 7.5 km radius of the proposed park. The foregone benefit net of inputs used is $91 per household per year. A comprehensive contingent valuation survey of the same villages, indicated that the willingness to pay (WTP) for access denied to the park area amounted to $108 per household per year.

A novel international travel cost (or recreation demand) model was used to determine the value of the proposed park to international tourists. The average tourist earned about $60 000 per year, had 15 years of education, and spent about $2900 per trip. Two empirical models— random utility (RU) and typical trip (TT)—were used to measure value, yielding estimates of $24 and $45 per trip. A separate contingent value survey of eco-tourists yielded a mean willingness-to-pay of $65 per trip.

Table 2.2: Value of Agricultural and Forestry Activities
(*Source:* Kramer *et al.*, 1995).

Activity	Number of observations	Total annual value for all villages (US$)	Annual mean value per household (US$)
Rice	351	44 928	128
Fuelwood	316	13 289	38
Crayfish	19	220	12
Crab	110	402	3.7
Tenreck	21	125	6
Frog	11	71	6.5

Conclusions: All these results, and the total present value of benefits from these alternative uses of the rainforest (by local villagers or tourists) are summarized in Table 2.3. Several tentative conclusions can be drawn from the results of this study. Non-market valuation techniques can provide useful information for economic evaluation of national parks. A major strength of this study is the opportunity to compare valuation techniques. For the village component, the estimated benefits from park use based on two entirely different methods, opportunity cost analysis and contingent valuation method, were remarkably similar ($91 and $108 per household per year). The estimates of tourist benefits based on the travel cost method and contingent valuation method were somewhat more disparate ($24 to $65 per trip) but it is noteworthy that the benefit estimates are of the same order of magnitude. We note that the higher contingent valuation estimate may reflect some non-use values, while the recreation demand method is mainly for use value only.

This type of analysis would have implications for policy, investment decisions, resource mobilization, and project design and management. It can help governments to decide how to (a) allocate scarce capital resources among competing land-use activities; (b) choose and implement investments for natural resource conservation and development; (c) determine pricing, land use, and incentive policies; (d) determine compensation for local villagers for foregone access to forest areas designated as national parks; and (e) value the park as a global environmental asset to foreigners (thus attracting external assistance for conservation programmes at the local level).

At the same time, the findings indicate future issues. Reliance on WTP is fundamental to the economic approach, but tends to overemphasize the importance of value ascribed to richer foreign visitors. Assuming mutually exclusive alternative uses of the park, the costs (represented by the foregone benefits of villagers) are significantly less than potential benefits to tourists. If conflicting claims to park access were to be determined purely on this basis, residents (especially, the poor local villagers) are more likely to be excluded. Therefore, the socio-cultural concepts of sustainable development (especially intra-generational equity and distributional concerns) would need to be invoked to protect the basic rights of local residents—for example, in the form of a 'safe minimum' degree of access to park facilities, irrespective of WTP-based benefits that are dependent on income levels.

Table 2.3: Summary of Economic Analysis of Mantadia National Park.

Estimates of Welfare Losses to Local Villagers from Establishment of Park		
Method used	Annual mean value per household	Total present value[a]
Opportunity cost	(US $) 91	(US $) 673 078
Contingent valuation	108	566 070
Estimates of Welfare Gains to Foreign Tourists from Establishment of Park		
Method used	Annual mean value per trip	Total present value[a]
Recreation Demand 1 (RU)	(US $) 24	(US $) 936 000
Recreation Demand 2 (TT)	45	1 750 000
Contingent Valuation	65	2 530 000

[a] Discount rate = 10%.

2.3.4 Local-project Scale: Fuelwood Stoves and Hydroelectric Power

The procedures for conventional environmental and social assessment at the project/local level (which are now well accepted world wide), may be readily adapted to assess the environmental and social effects of micro-level activities (World Bank, 1998)—see also Figure 2.5. The OECD (1994) has pioneered the 'Pressure-State-Response' framework to trace socioeconomic-environment linkages. This P-S-R approach begins with the pressure (*e.g.* population growth), then seeks to determine the state of the environment (*e.g.* ambient pollutant concentration), and ends by identifying the policy response (*e.g.* pollution taxes). Specific methods for economic valuation of environmental and social impacts were described earlier in Box 2.3. The practical application of such techniques at the local level were illustrated in the previous case study. When valuation is not feasible for certain impacts, MCA may be used.

2.3.4.1 Case Study 7: Multicriteria Analysis of a Fuelwood Stove Project

Figure 2.11 illustrates how an MCA-based analysis at the project level could provide balanced treatment of economic, social and environmental considerations. The stylized project evaluation involves the case of an improved fuelwood burning stove.

As discussed earlier, MCA offers policy-makers an alternative when progress toward multiple objectives cannot be measured in terms of a single criterion (*e.g.* monetary values). Take the case of an efficient

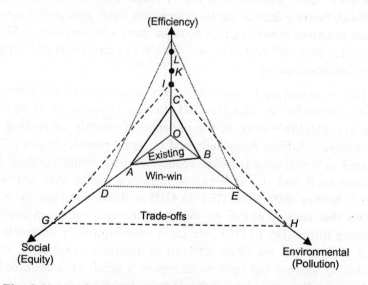

Fig. 2.11: Analysing the sustainability of an improved fuelwood stove using multicriteria analysis

Source: Adapted from Munasinghe [1993]

fuelwood stove—an end-use option for sustainable energy development. While the economic value of such a cookstove is measurable, its contribution to social and environmental goals is not easily valued in economic terms. As shown in Figure 2.11, outward movements along the axes trace improvements in three indicators: economic efficiency (net monetary benefits), social equity (improved health of poor energy users), and environmental pollution (reduced deforestation and GHG emissions).

We may assess the policy options as follows. First, triangle ABC represents the existing method of burning fuelwood (typically placing the cooking pot on three bricks). In this case, the indicators of economic efficiency, social equity, and overall environmental impact are all bad, because the stove uses fuelwood inefficiently, increases smoke inhalation (especially by women and children in poor households), and worsens GHG emissions and pressure on forest resources. Next, triangle DEF indicates a 'win-win' future option based on an improved fuelwood stove, in which all three indices improve. The economic gains would include monetary savings from reduced fuelwood use and increased productivity from reductions in acute respiratory infections, lung disease and cancer caused by pollutants in biomass smoke. Social gains would accrue from the fact that the rural poor benefit the most from this innovation—for example, due to the lighter health and labour burden on women and children, and the reduced time spent on collecting fuelwood, thereby increasing time spent on other productive activities. The environment benefits occur because more efficient use of fuelwood will reduce both deforestation and greenhouse emissions resulting from inefficient combustion.

After realizing such 'win-win' gains, other available options would require trade-offs. In triangle GIH, further environmental and social gains are attainable only at the expense of sharply increasing costs. For example, shifting from fuelwood to liquid petroleum gas (LPG) or kerosene as a fuel may increase economic costs, while yielding further environmental and social benefits. A policy-maker may not wish to make a further shift from DEF to GIH without knowing the relative weights that society places on the three indices—in sharp contrast to the move from ABC to DEF, which is unambiguously desirable. Such social preferences are often difficult to determine explicitly, but it is possible to narrow the options. Suppose a small economic cost, FL, yields the full social gain DG, while a large economic cost, LI, is required to realize the environmental benefit EH. Here, the social gain may better justify the economic sacrifice. Further, suppose that budgetary constraints limit costs to less than FK (where FL < FK < LI). Then, sufficient funds exist only to pay for the social benefits, and the environmental improvements will have to be delayed.

2.3.4.2 Case Study 8: Comparison of Hydroelectric Power Projects

In this case study, MCA is used to compare hydroelectric power schemes (for details, see Morimoto and Munasinghe, 2000). The three main sustainable development issues that are considered comprise the economic costs of power generation, ecological costs of biodiversity loss, and social costs of resettlement.

The principal objective is to generate additional kilowatt-hours (kWh) of electricity to meet the growing demand for power in Sri Lanka. As explained earlier in the section on cost-benefit analysis (CBA), we assume that the benefits from each additional kWh are the same. Therefore, the analysis seeks to minimize the economic, social and environmental costs of generating one unit of electricity from different hydropower sites. Following the MCA approach, environmental and social impacts are measured in different (non-monetary) units, instead of attempting to economically value and incorporate them within the single-valued CBA framework.

Environmental, Social and Economic Indicators: Sri Lanka has many varieties of fauna and flora, many of which are endemic or endangered. Often large hydro projects destroy wildlife at the dam sites and the downstream areas. Hence, biodiversity loss was used as the main ecological objective. The biodiversity index described in Box 2.4 was estimated for each hydroelectric site.

Although dam sites are usually in less densely populated rural areas, resettlement is still a serious problem in most cases. In general, people are relocated from the wet to the dry zone where soils are less rich, and therefore the same level of agricultural productivity cannot be maintained. In the wet zone, multiple crops including paddy, tobacco, coconuts, mangos, onions, and chilies can be grown. However, these crops cannot be cultivated as successfully in the dry zone, due to limited access to water and poor soil quality. Living standards often become worse and several problems (like malnutrition) could occur. Moreover, other social issues such as erosion of community cohesion and psychological distress due to change in the living environment, might arise. Hence, minimising the number of people resettled due to dam construction is one important social objective.

The project costs are available for each site, from which the critical economic indicator—average cost per kWh per year—may be estimated (for details, see Ceylon Electricity Board (CEB), 1987, 1988, 1989). The annual energy generation potential at the various sites ranges from about 11 to 210 KWh (see Figure 2.12). All three variables, the biodiversity index, number of people resettled, and generation costs, are weighted by the inverse of the amount of electrical energy generated. This scaling removes the influence of project size and makes them more comparable.

Some Basic Results: A simple statistical analysis shows that, pairwise, there is little correlation between the quantity of electricity generated, average generation cost, number of people resettled, and biodiversity index.

From Figure 2.12, it is clear that on a per kWh per year basis, the project named AGRA003 has the highest biodiversity index, HEEN009 has the highest number of resettled people, and MAHA096 has the highest average generation cost. Some important comparisons may be made. For example, KALU075 is a relatively large project where the costs are low, whereas MAHA096 is a smaller scheme with much higher costs with respect to all three indices. Another simple observation is that a project like KELA071 fully dominates GING053, since the former is superior in terms of all three indicators. Similar pairwise comparisons between other projects may be needed.

A three-dimensional analysis of sustainable development indicators for these hydropower sites is provided in Figure 2.13, where the axes represent economic, ecological, and social objectives, respectively. The distance from the origin to each coordinate point can be seen, and the closer to the origin, the better is the project in terms of achieving these three objectives. This type of analysis gives policy-makers some idea about which project is more favourable from a sustainable energy development perspective. Suppose we arbitrarily give all the three objectives an equal weight. Then, each project may be ranked according to its absolute distance from the origin. For example, rank 1 is given to the one closest to the origin, rank 2 is to the the second closest and so on, as shown in Figure 2.13. On this ad-hoc overall basis, from a sustainable energy development perspective, the most favourable project is GING074 (rank 1), whereas the least favourable one is MAHA096 (rank 22).

Conclusions: The strength of this type of analysis is in helping policy-makers to compare project alternatives more easily and effectively. The simple graphical presentations are more readily comprehensible,

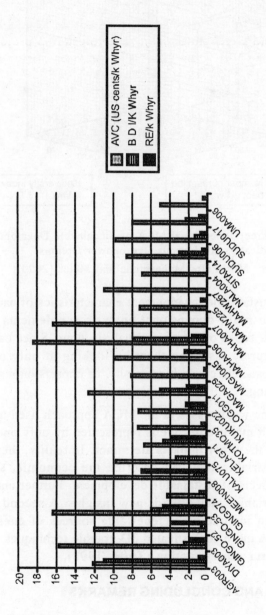

Fig. 2.12: Average Generation Costs (AVC), Biodiversity Index (BDI), and Number of Resettled People (RE) by Hydroelectric Project. All Indices are per kWh per year. Numbers of People Resettled and the Biodiversity Index are scaled for convenience (by the Multipliers 10^{-5} and 10^{-9} Respectively). The Values at the Top of the Graph Indicate the Annual Energy Generation in Gigawatt Hours (GWh).

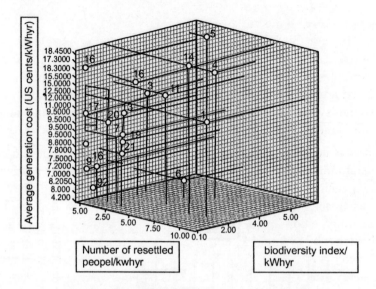

Fig. 2.13: Three Dimensional MCA of Sustainable Development
Indicators for Various Hydropower Options.
Source: Morimoto, Munasinghe and Meier [2000]

and identify the sustainable development characteristics of each scheme
quite clearly. The multi-dimensional analysis supplements the more
conventional CBA, based on economic analysis alone. Since each project
has different features, assessing them by looking at only one aspect
(*e.g.* generation costs, effects on biodiversity, or impacts on resettlement)
could be misleading.

There are some weaknesses in the MCA approach used here. First,
for simplicity each major objective is represented by only one variable,
assuming that all the other impacts are minor. In reality, there may be
more than one variable that can describe the economic, social and
environmental aspects of sustainable development. Further analysis that
includes other variables may provide new insights. A second extension
of this study is to include other renewable sources of energy in the
analysis. Finally, a more sophisticated 3D graphic techniques may yield
a better and clearer representation (Tufte, 1992).

2.4 SUMMARY AND CONCLUDING REMARKS

Sustainable development is one of the most important challenges
facing humankind in the 21st century. While no universally acceptable
practical definition exists as yet, the concept has evolved to encompass
three major points of view: economic, social and environmental. Each

viewpoint corresponds to a domain or system, which has its own distinct driving forces and objectives. The economic system is geared mainly towards improving human welfare (primarily through increases in the consumption of goods and services). The environmental domain focuses on protection of the integrity and resilience of ecological systems. The social system seeks to enrich human relationships and achieve individual and group aspirations.

There is no single overarching framework for sustainable development, but sustainomics attempts to describe 'a trans-disciplinary, integrative, balanced, heuristic and practical meta-framework for making development more sustainable'. It seeks to synthesize key elements from core disciplines like ecology, economics, and sociology, as well as others such as anthropology, biotechnology, botany, chemistry, demography, engineering, ethics, geology, information technology, philosophy, physics, psychology, and zoology. Methods that cross the economy-society-environment interfaces are also important, including environmental and resource economics, ecological economics, conservation ecology, social capital and inclusion, energetics and energy economics, sociological economics, environmental sociology, cultural economics, economics of sociology, and sociology of the environment. While building on earlier work, sustainomics constitutes a more neutral expression which focuses attention explicitly on sustainable development, and especially issues of concern to the developing world.

Comprehensiveness is an important requirement because sustainable development involves every aspect of human activity and involves complex interactions among socioeconomic, ecological and physical systems. The scope of analysis needs to extend from the global to the local scale, cover time spans extending to centuries (for example, in the case of climate change), and deal with problems of uncertainty, irreversibility, and non-linearity. The approach must not only integrate the economic, social and environmental dimensions of sustainable development, as well as related methodologies and paradigms in a consistent manner, but also provide balanced treatment of all these elements. Balance is also needed in the relative emphasis placed on traditional development *versus* sustainability. No single discipline could cope with the multiplicity of issues involved, and therefore a trans-disciplinary framework is required which would address the many facets, from concept to actual practice. Furthermore, the precise definition of sustainable development remains an elusive (and perhaps unreachable) goal. Thus, the less ambitious strategy of simply seeking to make development more sustainable, might offer greater promise. Such an incremental (or gradient-based) method is more practical, because many unsustainable activities are often easier to recognize and eliminate.

Although the current state of knowledge makes it rather difficult to provide a complete definition of sustainomics, this paper has identified some of its key constituent elements and how they might fit together. The basic intention was to sketch out preliminary ideas which would help to stimulate discussion and encouraging further contributions that are needed to flesh out the initial framework.

The environmental, social and economic criteria for sustainability play an important role in the sustainomics framework. The environmental interpretation of sustainability focuses on the overall viability and health of ecological systems—defined in terms of a comprehensive, multiscale, dynamic, hierarchical measure of resilience, vigour and organization. Natural resource degradation, pollution and loss of biodiversity are detrimental because they increase vulnerability, undermine system health, and reduce resilience. The notion of a safe threshold (and the related concept of carrying capacity) are important—often to avoid catastrophic ecosystem collapse. The nested hierarchy of ecological and social systems across scales and their adaptive cycles constitute a 'panarchy'. A system at a given level is able to operate in its stable (sustainable) mode, because of the continuity provided by the slower and more conservative changes in the super-system above it, while being simultaneously invigorated and energized by the faster cycles of change taking place in the sub-systems below it.

Social sustainability seeks to reduce the vulnerability and maintain the health (*i.e.* resilience, vigour and organization) of social and cultural systems, and their ability to withstand shocks. Enhancing human capital (through education) and strengthening social values and institutions (like trust and behavioural norms) are key aspects. Weakening social values, institutions and equity will reduce the resilience of social systems and undermine governance. Preserving cultural diversity and cultural capital across the globe, strengthening social cohesion and networks of relationships, and reducing destructive conflicts, are integral elements of this approach. In summary, for both ecological and socioeconomic systems, the emphasis is on improving system health and their dynamic ability to adapt to change across a range of spatial and temporal scales, rather than the conservation of some 'ideal' static state.

The modern concept underlying economic sustainability seeks to maximize the flow of income that could be generated while at least maintaining the stock of assets (or capital), which yield these beneficial outputs. Economic efficiency plays a key role—in ensuring both efficient allocation of resources in production, and efficient consumption choices that maximize utility. Problems of interpretation arise in identifying the kinds of capital to be maintained (for example, manufactured, natural, human and social capital stocks have been identified) and their substitutability. Often, it is difficult to value these assets and the services they provide, particularly in the case of ecological and social resources. The issues of uncertainty, irreversibility and catastrophic collapse pose additional difficulties, in determining dynamically efficient development paths.

Equity and poverty play an important role in the sustainomics framework. Both issues have not only economic, but also social and environmental dimensions, and therefore, they need to be assessed using a more comprehensive set of indicators (rather than income distribution alone).

Several analytical techniques have sought to provide integrated and balanced treatment of the economic, social and environmental viewpoints. If material growth is the main issue, while uncertainty is not a serious problem, and relevant data are available, then the focus is more likely to be on optimizing economic output, subject to (secondary) constraints that ensure social and environmental sustainability. Alternatively, if sustainability is the primary objective, conditions are chaotic, and data are rather weak, then the emphasis would be on paths which are economically, socially and environmentally durable or resilient, but not necessarily growth optimizing. Sustainomics attempts to use both optimal and durable approaches, by developing their potential to yield consistent and complementary results. In the same vein, sustainomics could also better reconcile the natural science view which relies more on flows of energy and matter, with the sociological and economic approaches that focus on human activities and behaviour. One potential area of application of sustainomics involves integrated assessment models, which contain a variety of submodels that represent ecological, geophysical and socioeconomic systems. Cost-benefit analysis and multi-criteria analysis are useful tools for analysing sustainable development issues.

The sustainomics framework would encourage crucial changes in the mindset of decision-makers, by helping them to focus on the structure of development, rather than just the magnitude of economic growth (conventionally measured). This process would make development more sustainable, through the adoption of environmentally- and socially-friendly strategies that enable us to use natural resource inputs more frugally and efficiently, reduce polluting emissions, and facilitate public participation in social decisions. Sustainomics serves as an essential bridge between the traditional techniques of decision making and modern environmental and social analysis, by helping to incorporate ecological and social concerns into the decision making framework of human society. Operationally, it plays this bridging role by helping to map the results of environmental and social assessments (EA and SA) onto the framework of conventional economic analysis of projects. Thus, sustainomics identifies practical social and natural resource management options that facilitate sustainable development.

The paper also illustrates these concepts, by applying them to case studies involving energy problems across the full range of spatial scales. At the global-transnational level, the first case study examines the interplay of optimality and durability in determining appropriate global GHG emission target levels, while the second explores methods of combining efficiency and equity to facilitate South-North cooperation for climate change mitigation. At the level of national-economy policies, the third case study describes how the action impact matrix may be used for policy analysis, while the fourth sets out approaches for restructuring growth to make long term development more sustainable. On the subnational-sectoral scale, the fifth case outlines methods for improving energy sector decision making in Sri Lanka, and the sixth examines rainforest management in Madagascar. Finally, at the project-local level, multi-criteria analysis is applied to the case of a fuelwood stove project, and to compare small hydroelectric power projects, using relevant economic, social and environmental indicators.

References and notes

Adger, W.N. (1999) 'Social vulnerability to climate change and extremes in coastal Vietnam', *World Development,* February, pp. 1–21.
Adriaanse, A. (1993) *Environmental Policy Performance Indicators*, Sdu, Den Haag.
Alfsen, K. H. and Saebo, H.V. (1993) 'Environmental quality indicators: background, principles and examples from Norway', *Environmental and Resource Economics*, October, Vol. 3, pp. 415–35.
Andersen, E. (1993) *Values in Ethics and Economics*, Harvard University Press, Cambridge MA.
Arrow, K.J., Cline, W., Maler, K.G., Munasinghe, M. and Stiglitz, J. (1995) 'Intertemporal equity, discounting, and economic efficiency', in *Global Climate Change: Economic and Policy Issues*, M. Munasinghe (Editor) World Bank, Washington DC.
Atkinson, G., Dubourg R., Hamilton, K., Munasinghe, M., Pearce, D.W. and Young, C. (1997) *Measuring Sustainable Development: Macroeconomics*

and the Environment, Edward Elgar, Cheltenham, UK.

Azar, C., Homberg, J. and Lindgren, K. (1996) 'Socio-ecological indicators for sustainability', *Ecological Economics*, Vol. 18, August, pp. 89–112.

Banuri, T. (1998) 'Human and environmental security', *Policy Matters*, Vol. 3, Autumn.

Banuri, T., Hyden, G., Juma, C. and Rivera, M. (1994) *Sustainable Human Development: From Concept To Operation: A Guide For The Practitioner*, UNDP, New York.

Bennet, R. (2000) 'Risky business', *Science News*, Vol. 158, pp. 190–1, September.

Bergstrom, S. (1993) 'Value standards in sub-sustainable development: On limits of ecological economics', *Ecological Economics*, Vol. 7, pp. 1–18, February.

Bohle, H.G., Downing, T.E. and Watts, M.J. (1994) 'Climate change and social vulnerability: toward a sociology and geography of food insecurity', *Global Environmental Change*, Vol. 4, No. 1, pp. 37–48.

Brown, P.G. (1998) 'Towards an economics of stewardship: the case of climate', *Ecological Economics*, Vol. 26, pp. 11–21, July.

Burton, I. (1997) 'Vulnerability and adaptive response in the context of climate and climate change', *Climatic Change*, Vol. 36, Nos. 1–2, pp. 185–196.

Ceylon Electricity Board (CEB) (1987) *Masterplan for electricity supply in Sri Lanka, Vol 1*, CEB, Colombo, Sri Lanka.

Ceylon Electricity Board (CEB) (1988) *Masterplan for electricity supply in Sri Lanka, Vol 2*, CEB, Colombo, Sri Lanka.

Ceylon Electricity Board (CEB) (1999) *Long Term Generation Expansion Plan 1999-2013*, CEB, Colombo, Sri Lanka.

Chambers, R. (1989) 'Vulnerability, coping and policy', *IDS Bulletin*, Vol. 20, No. 2, pp. 1–7.

Chenery, H. and Srinivasan, T.N. (Editors) (1988) (1989) *Handbook of Development Economics*, i and ii, North-Holland, Amsterdam.

Chichilnisky, G. and Heal, G. (Editors) (2000) *Environmental Markets: Equity and Efficiency*, Columbia University Press, New York, USA.

Cialdini, R.B. (2001) *Influence: Science and Practice*, Fourth Edition, Allyn and Bacon, London.

Clark, W.C. (1998) 'Visions of the 21st century: Conventional wisdom and other surprises in the global interactions of population, technology and environment', in K. Newton, T. Schweitzer, J. P. Voyer (Editors), *Perspective 2000: Proceedings of a conference sponsored by the Economic Council of Canada, December*, Economic Council of Canada, Ottawa, pp. 7-32.

Clarke, C. and Munasinghe, M. (1995) 'Economic aspects of disasters and sustainable development', in M. Munasinghe and C. Clarke (Editors) *Disaster Prevention for Sustainable Development*, Int. Decade of Natural Disaster Reduction (IDNDR) and World Bank, Geneva and Wash. DC.

Colding, J., and Folke, C. (1997) 'The relations among threatened species, their protection, and taboos', *Conservation Ecology*, Vol. 1, No. 1, p. 6. Available from: http://www.consecol.org/vol1/iss1/art6.

Cole, M.A., Rayner, A.J. and Bates, J.M. (1997) 'Environmental quality and economic growth', *University of Nottingham, Department of Economics Discussion Paper; 96/20*, pp.1–33, December.

Coleman, J. (1990) *Foundations of Social Theory*, Harvard Univ. Press, Cambridge, MA.

Commission on Sustainable Development (1998) *Indicators of Sustainable Development*, New York.

Conservation Ecology (various issues), published electronically; URL: http://www.consecol.org.

Costanza, R. (2000) 'Ecological sustainability, indicators and climate change', in M. Munasinghe and R. Swart (Editors) *Climate Change and its Linkages with Development, Equity and Sustainability*, IPCC, Geneva, Switzerland.

Costanza, R., Cumberland J., Daly, H., Goodland, R. and Norgaard, R. (1997) *An Introduction to Ecological Economics*, St. Lucia's Press, Boca Raton FL, USA.

Daly, H.E. and Cobb, J.B. Jr. (1989) *For the Common Good*, Beacon Press, Boston MA.

Dasgupta, P. and Maler, K.G. (1997) 'The resource basis of production and consumption: an economic analysis', in P. Dasgupta and K.G. Maler (Editors) *The Environment and Emerging Development Issues, Vol. 1*, Claredon Press, Oxford, UK.

Dreze, J. and Sen, A. (1990) *Hunger and Public Action*, Clarendon Press, Oxford.

Ecological Economics (2000) *Special Issue on the Human Actor in Ecological-Economic Models*, Vol.35, No.3, December.

Ecological Economics (various issues), Elsevier, Amsterdam.

Environmental Ethics (various issues), Elsevier, Amsterdam.

Faucheux, S., Pearce, D. and Proops, J. (Editors) (1996) *Models of Sustainable Development*, Edward Elgar Publ., Cheltenham, UK.

Fisher I. 1906 (reprinted 1965) *The Nature of Capital and Income*, Augustus M. Kelly, New York NY, USA.

Freeman, A.M., (1993) *The Measurement of Environmental and Resource Values: Theory and Methods*, Resources for the Future, Washington DC.

Georgescu-Roegen, N. (1971) *The Entropy Law and the Economic Process*, Harvard Univ. Press, Cambridge, MA, USA.

Gilbert, A. and Feenstra, J. (1994) 'Sustainability indicators for the Dutch environmental policy theme 'diffusion' cadmium accumulation in soil', *Ecological Economics*, Vol. 9, pp. 253–65, April.

Gintis, H. (2000) 'Beyond *homo economicus*: evidence from experimental economics', *Ecological Economics*, Vol. 35, No. 3, pp. 311-23.

Githinji, M. and Perrings, C. (1992) 'Social and ecological sustainability in the use of biotic resources in sub-Saharan Africa: rural institutions and decision making in Kenya and Botswana', *Mimeo.*, Beijer Institute and University of California, Riverside, July.

Grootaert, C. (1998) 'Social capital: the missing link', *Social Capital Initiative Working Paper No. 3*, World Bank, Washington DC.

Gunderson, L., and Holling, C.S. (2001) *Panarchy: understanding transformations in human and natural systems.* Island Press, New York.

Hall, C. (Editor). (1995) *Maximum Power: The Ideas and Applications of H.T. Odum,* Colorado Univ. Press, Niwot, CO, USA.

Hanna, S., and Munasinghe, M. (1995) *Property Rights in Social and Ecological Context,* Beijer Institute and the World Bank, Stockholm and Washington D.C.

Hicks, J. (1946) *Value and Capital,* 2nd edition, Oxford University Press, Oxford, UK.

Holling, C.S. (1973) 'Resilience and stability of ecological systems', *Annual Review of Ecology and Systematics,* Vol. 4, pp. 1–23.

Holling, C.S. (1986) 'The resilience of terrestrial ecosystems: local surprises and global change', in W.C.Clark and R.E.Munn (Editors) *Sustainable Development of the Biosphere,* Cambridge University Press, Cambridge, UK, pp. 292-317.

Holmberg, J., and Karlsson, S. (1992) 'On designing socio-ecological indicators', in U. Svedin and Bhagerhall-Aniansson (Editors) *Society and Environment: A Swedish Research Perspective,* Kluwer Academic, Boston.

Holtz-Eakin, D. and Selden, T.M. (1995) 'Stoking the fires? CO_2 emissions and economic growth', *National Bureau of Economic Research, Working Paper Series, 4248,* pp. 1–38. December

IPCC (1996a) *Climate Change 1995: Economic and Social Dimensions of Climate Change,* J.P. Bruce, *et al.,* (Editors) Cambridge University Press, London..

IPCC (1996b) *Climate Change 1995: Impacts, Adaptations and Mitigation of Climate Change,* Watson, R.T. *et al.,* (Editors) Cambridge University Press, London.

IPCC (1997) *Climate Change and Integrated Assessment Models (IAMs),* Geneva, Switzerland.

IPCC (1999) *Special Report on Technology Transfer,* draft report, IPCC, Geneva.

Islam, Sardar M.N. (2001) 'Ecology and optimal economic growth: an optimal ecological economic growth model and its sustainability implications', in M. Munasinghe, O. Sunkel and C. de Miguel (Editors) *The Sustainability of Long Term Growth,* Edward Elgar Publ., Cheltenham, UK.

Jepma, C. and Munasinghe, M. (1998) *Climate Change Policy,* Cambridge Univ. Press, Cambridge, UK.

Kramer, R.A., Sharma N. and Munasinghe, M. (1995) *Valuing Tropical Forests,* The World Bank, Washington DC, USA.

Kuik, O. and Verbruggen, H. (Editors) (1991) *In Search of Indicators of Sustainable Development,* Kluwer, Boston.

Kverndokk, S. (1995) 'Tradeable CO_2 emission permits: initial distribution as a justice problem', *Environmental Values,* Vol. 4, pp. 129–48.

Liverman, D., Hanson, M., Brown, B.J. and Meredith, R. Jr. (1988) 'Global sustainability: towards measurement', *Environmental Management,* Vol. 12, pp. 133–143.

Lonergan, S.L. (1993) 'Impoverishment, population and environmental degradation:

the case for equity', *Environmental Conservation*, Vol. 20, No. 4, pp. 328–334.

Lovelock, L. (1975) *Gaia: A New Look at Life on Earth*.

Ludwig, D., Walker, B. and Holling, C.S. (1997) 'Sustainability, stability, and resilience'. *Conservation Ecology* (online) Vol. 1, No. 1, p.7. Available from. URL: http://www.consecol.org/vol 1/iss1/art7.

Maler, K.G. (1990) 'Economic theory and environmental degradation: a survey of some problems', *Revista de Analisis Economico*, Vol. 5, pp.7–17, November.

Maslow, A.H. (1970) *Motivation and Personality*, Harper and Row, New York.

Meier, P. and Munasinghe, M. (1994) *Incorporating Environmental Concerns Into Power Sector Decision Making*, The World Bank, Washington DC.

Moffat, I. (1994) 'On measuring sustainable development indicators', *International Journal of Sustainable Development and World Ecology*, Vol. 1, pp.97–109.

Morimoto, R. and Munasinghe, M. (2000) 'Sustainable energy development: assessing the economic, social and environmental implications of hydropower projects in Sri Lanka', *MIND Research-Discussion Paper No.2*, Munasinghe Institute For Development (MIND), Colombo, Sri Lanka.

Moser, C. (1998) 'The asset vulnerability framework: reassessing urban poverty reduction strategies', *World Development*, Vol.26, No.1, p.1-19.

Munasinghe, M. (Editor) (1989) *Computers and Informatics in Developing Countries*, Butterworths Press, London UK, for the Third World Academy of Sciences, Trieste, Italy.

Munasinghe, M. (1990) *Energy Analysis and Policy*, Butterworth-Heinemann, London, UK.

Munasinghe, M. (1993) *Environmental Economics and Sustainable Development*, World Bank, Washington, DC, USA.

Munasinghe, M. (1994) 'Sustainomics: a transdisciplinary framework for sustainable development', *Keynote Paper, Proc. 50th Anniversary Sessions of the Sri Lanka Assoc. for the Adv. of Science* (SLAAS), Colombo, Sri Lanka.

Munasinghe, M. (1995) 'Making growth more sustainable', *Ecological Economics*, Vol. 15, pp.121–4.

Munasinghe, M. (Editor) (1997) *Environmental Impacts of Macroeconomic and Sectoral Policies*, International Society for Ecological Economics and World Bank, Solomons, MD and Washington DC.

Munasinghe, M. (1998a) 'Climate change decision-making: science, policy and economics', *International Journal of Environment and Pollution*, Vol. 10, No. 2, pp. 188–239.

Munasinghe, M. (1998b) 'Countrywide policies and sustainable development: are the linkages perverse?', in T. Teitenberg and H. Folmer (Editors) *The International Yearbook of International and Resource Economics*, Edward Elgar Publ., London, UK.

Munasinghe, M. (1998c). 'Is environmental degradation an inevitable consequence of economic growth', *Ecological Economics*, December.

Munasinghe, M. (2000) 'Development, equity and sustainability in the context of climate change', *IPCC Guidance Paper*, Intergovernmental Panel on Climate Change, Geneva.

Munasinghe, M. and Cruz, W. (1994) *Economywide Policies and the Environment*, The World Bank, Washington DC, USA.

Munasinghe, M. and King, K. (1992) 'Accelerating ozone layer protection in developing countries', *World Development* Vol. 20, April, pp.609–18.

Munasinghe, M. and Munasinghe. S. (1993) 'Enhancing south-north cooperation to reduce global warming', *Paper presented at the IPCC Meeting on Global Warming*, Montreal, May.

Munasinghe, M. and Shearer, W. (Editors) (1995) *Defining and Measuring Sustainability: The Biogeophysical Foundations*, UN University and World Bank, Tokyo and Washington, DC.

Munasinghe, M., Sunkel, O. and de Miguel, C. (Editors) (2001) *The Sustainability of Long Term Growth*, Edward Elgar Publ., London, UK.

Munasinghe, M. and Swart, R. (Editors) (2000) *Climate Change and its Linkages with Development, Equity and Sustainability*, Intergovernmental Panel on Climate Change (IPCC), Geneva, Switzerland.

Narada, The Venerable (1988) *The Buddha and His Teachings,* Buddhist Missionary Society, Kuala Lumpur, Malaysia, Fourth Edition.

Nordhaus, W. and Tobin, J. (1972) 'Is growth obsolete?', *Economic Growth*, National Bureau of Economic Research (NBER), Columbia University Press, New York, NY, USA.

North, D. (1990) *Institutions, Institutional Change and Economic Performance*, Cambridge Univ. Press, Cambridge, UK.

OECD (1994) *Environmental Indicators*, OECD, Paris, France.

Olson, M. (1982) *The Rise and Decline of Nations*, Yale Univ. Press, New Haven, CN, USA.

Opschoor, H. and Reijnders, L. (1991) 'Towards sustainable development indicators', in O. Kuik and H. Verbruggen (Editors) *In Search of Indicators of Sustainable Development*, Kluwer, Boston, MA, USA.

Parris, T.M. and Kates, R.W. (2001) *Characterizing a Sustainability Transition: The International Consensus*, Research and Assessment Systems for Sustainability Discussion Paper, Environment and Natural Resources Program, Belfer Center for Science and International Affairs, Kennedy School of Government, Harvard University, Cambridge, MA, USA.

Pearce, D.W. and Turner, R.K. (1990) *Economics of Natural Resources and the Environment*, Harvester Wheatsheaf, London, UK.

Perrings, C. and Opschoor, H. (1994) *Environmental and Resource Economics,* Edward Elgar Publ., Cheltenham, UK.

Perrings, C., Maler, K.G. and Folke, C. (1995) *Biodiversity Loss: Economic and Ecological Issues*, Cambridge University Press, Cambridge, UK.

Petersen, G.D., Allen, C.R. and Holling, C.S. (1998) 'Diversity, ecological function, and scale: resilience within and across scales', *Ecosystems,* Vol. 1.

Pezzey, J. (1992) 'Sustainable development concepts: an economic analysis', *Environment Paper No. 2*, World Bank, Washington DC.

Pigou, A.C. (1932) *The Economics of Welfare*, Macmillan, London, UK

Pimm, S.L. (1991) *The Balance of Nature?*, University of Chicago Press, Chicago, Illinois, USA.

Putnam, R.D. (1993) *Making Democracy Work: Civic Traditions in Modern Italy*, Princeton Univ. Press, Princeton.

Rawls, J.A. (1971) *Theory of Justice*, Harvard Univ. Press, Cambridge MA, USA.

Rayner, S. and Malone, E. (Editors) (1998) *Human Choice and Climate Change*, pp. 1–4, Batelle Press, Columbus OH, USA.

Ribot, J.C., Najam, A. and Watson, G. (1996) 'Climate variation, vulnerability and sustainable development in the semi-arid tropics', in J.C Ribot, A.R. Magalhaes and S.S. Pangides (Editors) *Climate Variability, Climate Change and Social Vulnerability in the Semi-Arid Tropics*, Cambridge University Press, Cambridge, UK.

Robson, A.J. (2001) 'The biological basis of human behavior', *Journal of Economic Literature*, Vol. XXXIX, pp.11-33, March.

Ruitenbeek, H.J. (1996) 'Distribution of ecological entitlements: implications for economic security and population movement', *Ecological Economics*, Vol. 17, pp. 49–64.

Schutz, J. (1999) 'The value of systemic reasoning', *Ecological Economics*, Vol. 31, No. 1, pp. 23–29, October.

Sen, A.K. (1981) *Poverty and Famines: An Essay on Entitlement and Deprivation*, Clarendon, Oxford, UK.

Sen, A.K. (1984) *Resources, Values and Development*, Blackwell, Oxford, UK.

Sen, A.K. (1987) *On Ethics and Economics*, Basil Blackwell, Cambridge MA, USA.

Sengupta, R. (1996) *Economic Development and CO_2 Emissions*, Institute for Economic Development, Boston University, Boston MA.

Siebhuner, B. (2000) 'Homo sustinens—towards a new conception of humans for the science of sustainability', *Ecological Economics*, Vol. 32, pp. 15–25.

Solow, R. (1986) 'On the intergenerational allocation of natural resources', *Scandinavian Journal of Economics,* Vol. 88, No. 1, pp. 141–9.

Squire, L. and van der Tak, H. (1975) *Economic Analysis of Projects*, Johns Hopkins Univ. Press, Baltimore MD, USA.

Stern, N.H. (1989) 'The economics of development: a survey', *Economic Journal*, 99.

Teitenberg, T. (1992) *Environmental and Natural Resource Economics*, Harper Collins Publ., New York NY.

Tellus Institute (2001) *Halfway to the Future: Reflections on the Global Condition*, Tellus Institute, Boston. MA, USA.

Temple, J. (1999) 'The new growth evidence', *Journal of Econ. Literature*, Vol. XXXVII. pp. 112–54, March.

Toth, F. (1999) 'Decision analysis for climate change', in M. Munasinghe (Editor) *Climate Change and Its Linkages With Development, Equity and Sustainability*, Intergovernmental Panel on Climate Change (IPCC), Geneva, Switzerland.

Tufte, E.R. (1992) *The Visual Display of Quantitative Information*, Graphics Press, London UK.

UN (1993) *Agenda 21*, United Nations, New York.

UN (1996) *Indicators of Sustainable Development: Framework and Methodology*, New York.

UN Statistical Office (1993) *Integrated Environmental and Resource Accounting*, Series F, No. 61, United Nations, New York.

UNDP (1998) *Human Development Report*, New York.

UNEP, IUCN, and WWF (1991) *Caring for the Earth*, UNEP, Nairobi, Kenya.

UNFCCC (United Nations Framework Convention on Climate Change) (1993) *Framework Convention on Climate Change: Agenda 21*, United Nations, New York.

Unruh, G.C. and Moomaw, W.R. (1998) 'An alternative analysis of apparent EKC-type transitions', *Ecological Economics*, Vol. 25, pp. 221–229, May.

WCED (World Commission on Environment and Development) (1987) *Our Common Future*, Oxford University Press, Oxford, UK.

Westing, A. (1992) 'Environmental refugees: a growing category of displaced persons', *Environmental Conservation*, Vol. 19, No. 3, pp. 201–207.

Westra, L. (1994) *An Environmental Proposal for Ethics: The Principle of Integrity*, Rowman and Littlefield, Lanham MA, USA.

World Bank (1997) 'Expanding the measures of wealth: indicators of environmentally sustainable development', Environment Department, World Bank, Washington DC, USA.

World Bank (1998) *Environmental Assessment Operational Directive* (EAOD4.01), Washington DC, USA.

3

Valuation as Part of a Microeconomics for Ecological Sustainability[1]

Peter Söderbaum

Some actors in society are more powerful than others. In their thinking and practice, all these actors refer to specific images of man, of business, of markets and so on. Where do these ideas and images come from and what do they look like? Which images are most influential? John Maynard Keynes argued that many important ideas stem from the work of economists:

> "The ideas of economists and political philosophers . . . are more powerful than commonly understood. Indeed the world is ruled by little else. Practical men, who believe themselves to be quite exempt from any intellectual influences, are usually the slaves of some defunct economist." (from Fusfeld 1994, p. 1)

Examples of such ideas that play a role in public debate and in the decision making and practice of politicians, bureaucrats and businessmen is the mechanistic interpretation of markets in terms of supply and demand with associated beliefs in 'the invisible hand', a concept used by Adam Smith, but older in origin. Other examples of ideas that have become common include the view of firms as reducible to profit maximizing entities and of human beings as consumers who maximize utility.

But why is it that specific ideas get a strong hold on the thinking habits of many scientists and practical men? According to one view, economists and perhaps also the 'political philosophers' referred to by

---

[1] This chapter is a revised version of chapter 5 in O'Connor and Spash eds. 1999.

Keynes, seek truth and when successful in this endeavour, disseminate their knowledge to various actors in society and to the public at large. It is believed that the best knowledge will automatically gain acceptance in society.

In this paper, a second model referred to as an actor-network model, or alternatively, an interaction model will be emphasized. According to this view, models, theories and conceptual frameworks, especially in the social sciences, should not exclusively be seen as a matter of science but also of ideology. Concepts and theories are, at least in part, socially constructed to serve specific scientific and ideological purposes. Ideology is here defined in broad terms, as referring to 'ideas about means and ends'. The acceptance of specific conceptual simplifications, such as the aforementioned 'invisible hand', is as much a matter of ideology as science.

I believe that viewing interaction in society as a process of mutual learning, where science as well as ideology is involved, is a better representation of reality than the traditional 'truth dissemination' model. Also scientists learn as part of this interaction and they have themselves an ideological orientation and are not completely independent of the ongoing ideological debate in society. According to the very logic of the second model—which sees science and ideology as interconnected—this paper also reflects my own ideological ambition in searching for a microeconomics that will be helpful in dealing fruitfully with problems of environment and development.

Ecological economics is an interdisciplinary field of study which to a large extent is value driven (Cf. Krishnan *et al.* eds 1995). Ecological sustainability is widely espoused as an essential norm, and the ambition is to move the economy and society closer to this norm. Following in the wake of more than 20 years of activism, much of it by non-governmental organizations, the Rio conference in 1992 and Agenda 21 has stimulated work in organizations as well as municipalities to improve environmental performance. Ongoing activities are exemplified by Environmental Management Systems (EMS), such as ISO14 001, Environmental Auditing, and Environmental Labelling. Various tools such as Environmental Impact Assessment and Life Cycle Analysis have been developed and are recommended for use in individual countries and in the European Community. Reference is often made to Green Marketing, Green Consumerism, Greening of Business and so on. At issue in this paper is however whether neoclassical economics is a useful conceptual framework to understand and illuminate these phenomena or whether other theories and models more in line with social sciences such as sociology, political science, educational science, management science etc. have something to offer.

In what follows, I will not question that different parts of neoclassical microeconomics can be used in attempts to deal with environmental problems, but I will at the same time argue that it would be unwise to rely exclusively on the neoclassical approach. Neoclassical microeconomics was mainly constructed for other purposes, for instance monetary and financial policy at the national level. It should perhaps not be expected that it is at the same time 'optimal' or even 'satisficing' for environmental management purposes. Other theoretical perspectives in economics, for instance, institutionalism, and other social sciences appear to have a lot to offer to facilitate understanding of the Greening phenomena referred to and to contribute to the further development of tools and practices (Söderbaum 2000).

Essential components in neoclassical microeconomics are a view of man, usually referred to as Economic Man, a view of organizations, usually seen as firms, a view of markets in terms of supply and demand, a view of decision making and valuation in terms of monetary Cost-Benefit Analysis (CBA), and a view of social change in terms of Public Choice Theory. The idea here is not to find alternative views for each phenomenon with the promise to do the same job, but better. What is proposed is rather a conceptual framework that partly focuses on different phenomena and which can do a different job from that usually attributed to neoclassical economics.

The alternative view to be presented includes the following elements:

- Man is seen as a socio-psychological and political being and referred to as Political Economic Person. This political being is assumed to be led by a specific valuational or 'ideological orientation' in his or her behaviour. He or she can be seen as an 'actor'.

- Actors engage in activities and are connected in relationships and networks. Relationships may be of a non-market or market kind. Interaction outside markets, for instance public dialogue, may be as important for social and economic change as market behaviour.

- Markets, organizations and social change are interpreted in actor-network terms.

- A holistic view of economics and efficiency. 'Holistic' here refers to a multidimensional (and disaggregated) idea of resources and impacts, as well as an analysis in terms of patterns and profiles (cf. Figure 3.3). A holistic idea of economics (efficiency) is then seen as opposed to one-dimensional reductionism, for instance the 'monetary reductionism' of Cost-Benefit Analysis.

- A view of valuation and rationality as being a matter of 'ideological orientation' and a disaggregated approach to decision-making in business and at the societal level.

3.1 POLITICAL ECONOMIC PERSON

There is a political element in all kinds of social science, economics included. No paradigm and no research project can claim value neutrality in any final sense. "Valuations are always with us" as argued by Gunnar Myrdal (1978 p. 778).

"Disinterested research there has never been and can never be. Prior to answers there must be questions. In the questions raised and the viewpoint chosen, valuations are implied." (*ibid.* p. 779).

I will here go one step further by arguing that the political element is an important aspect of all human roles and all behaviour and thus not limited to the professional role as scholar in economics or other social sciences. It is assumed that man is a political being and to make this clear it will be suggested that Economic Man is replaced by a Political Economic Person (PEP). Or, to put it in the vocabulary of the actor-network model, man is seen as an actor who is embedded in a web of social relationships.

Figure 3.1 is an attempt to singleout essential concepts for an understanding of human beings as actors in society. An individual can be described in terms of 'roles', 'motives', and 'relationships', and 'activities' as 'basic concepts' and 'identity', 'ideological orientation', 'network', 'activity pattern', 'resources' and 'power' as more 'integrative concepts'. The individual interacts over time with her specific context which again can be understood as 'social', 'cultural', 'institutional', 'political', 'physical', 'man-made' and 'ecological'. These aspects of the context should not be seen as being easily separated but overlap in various degrees.

The Political Economic Person model reflects more of the complexities of human life than the Economic Man model. The latter focuses on and is limited to roles, motives, relationships, and activities connected with markets while PEP broadens the scope to include non-market roles, non-market motives, non-market relationships and activities. Different roles such as citizen, professional, consumer, wage-earner etc. are integrated in a person's 'identity' and 'ideological orientation'. The individual's relationships are part of networks (of relationships) and her activities form a pattern that can be referred to as a life-style. Resources of various kinds and the power position of the person will either extend or limit a person's possibilities for action and more generally for adaptation to her context (Cf. arrows in Figure 3.1). The adaptation process is furthermore characterized by stability and inertia as well as change. The individual is learning and 'moving' in a context that is changing over time.

From the point of view of ecological economics, it is essential to note that individuals—much like the economy as a whole—is embedded in a social and ecological context. Our life-style will have impacts of one kind or another upon other people and ecosystems. In understanding humans as actors in the economy and society, various kinds of roles, motives, relationships and activities—rather than exclusively those that are connected with markets—are assumed to be relevant. The individual is seen as a social being guided by an 'ideological orientation'. The total set of motives or interests combine to such a valuational or ideological orientation. Even 'world view' or 'Weltanschauung' is a

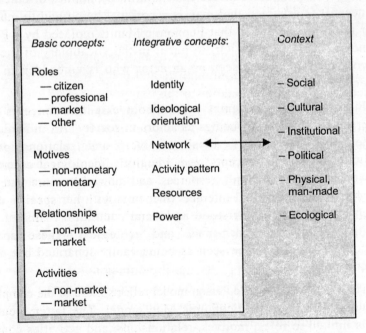

Fig. 3.1: Political Economic Person. Essential Concepts for an Understanding of a Person as Actor Adapting to her Context.

relevant description of this 'holistic' orientation, comprising cognitive as well as valuational elements. Inspite of tensions between various motives and interests, the individual is somehow held together through ideas of his or her role or identity in relation to each specific socio-cultural context. Dissonance theory, learning theories and other parts of social psychology are seen as relevant and useful in understanding behaviour. The individual strives for some congruence and balance between roles, activities and interests, and may experience such balance, but incongruence and tensions are equally characteristics of the human existence.

Egoistic versus other-related (or community oriented) motives are an example of such tensions. This points to a view of man as a moral being where responsibility in relation to others and society at large becomes a potential issue. Amitai Etzioni, for instance, has propounded an 'I &We Paradigm' (Etzioni 1988). According to him, the fact that there is a strong ego in each healthy individual is not sufficient reason to denigrate or exclude the social and ethical aspects of human life. Each individual plays a part in many groups, *i.e.* 'we-contexts' and such relationships involve a number of tensions and ethical issues. (Ethics is, of course, also relevant for 'we-they' relationships, for instance in situations of conflict.) Similarly Amartya Sen, an open-minded, mainly neoclassical economist, has argued in favour of explicit consideration of ethics in economics (1987).

In their models, neoclassicists tend to see individuals as robot-like, instant optimizers. Institutionalists and many representatives of other social sciences tend to point to the important role of habits in human behaviour. The individual is largely 'locked into' specific habits of thought and specific habitual activities that together form a pattern that can be referred to as a life-style. Herbert Simon´s early arguments about selective perception, limited cognitive capacity and search costs are relevant here (Simon, 1945). As humans we tend to stick to familiar environments and use various rules of thumb to deal with complexity. The development of a habit can be expressed in terms of increases in the probability of a specific behavior (like purchasing a specific brand A of coffee among available alternatives A, B and C) for successive trials, *i.e.* purchasing situations of a similar kind (cf. Howard, 1963). Emphasis on habitual behaviour does not exclude the possibility of 'problemistic search' and conscious decision making. At times the individual perceives a problem and identifies alternative courses of action. Habits are reconsidered and behaviour may change. Such decision situations can be discussed in conventional terms of maximizing an objective function, subject to various constraints. In what follows, a more holistic idea of rationality, related to the ideological orientation of an actor, will be emphasized.

The theory of the consumer as part of neoclassical microeconomics is of some interest when discussing environmental policy issues, for instance expected impacts of eco-taxes. Neoclassical public choice theory is similarly useful for understanding possible behaviour of individuals in professional roles. But a more holistic attempt to integrate various human roles seems warranted. The theory of the consumer is limited not only in the sense that one human role is emphasized at the expense of all others. In addition, consumer tastes or preferences are taken as given. As part of an imagined value-neutrality, the neoclassical scholar regards it as external to his/her role to problematize the values and life-styles of consumers. But if, as many suggest, environmental problems are connected with present consumer tastes and life-styles and more generally with dominating world-views in industrialized countries, then the neoclassical approach implies that essential aspects of the problems faced are overlooked. Focusing instead on Political Economic Person and 'ideological orientation' means that the different consumer preferences and life-styles of two individuals are no longer regarded as equally justified. Supported by a simultaneous, facilitating public policy or not, individuals may move in a step-by-step manner away from life-styles that are environmentally destructive toward those that are environmentally more beneficial. But again, whether such moves represent an advance is a matter of ethics and ideological orientation of the observer.

Political Economic Person assumptions will play a key role throughout the present essay. Reference will be made to political economic relationships, political economic networks, political economic organizations or management regimes, and political economic valuation of activities, projects or policies on the basis of ideological orientation.

3.2 RELATIONSHIPS AND NETWORKS

Individuals with their roles, activities and ideological orientation relate to each other as suggested by Figure 3.2. The description of each actor is simplified when compared with the one of Figure 3.1. Resources and power are regarded as essential in understanding interaction. Knowledge is regarded as an essential part of the resources and power of an actor as is the case with control over property of various kinds, e.g. financial and physical property.

All kinds of relationships between actors are potentially of importance in understanding environmental policy issues and environmental performance. Public and private debate may be as relevant as market relationships or relationships within an organization. And not all organizations are business companies. Some organizations are universities; others can be described as environmental organizations. What goes on at universities, or among intellectuals more generally, may be as decisive for the sustainability issue as anything else.

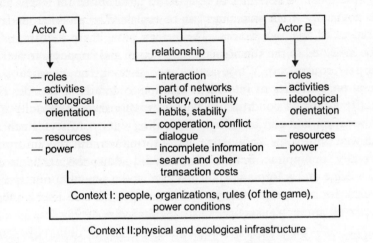

Fig. 3.2: Essential Aspects of a Relationship Between Two Actors.

The importance of public and private debate is emphasized here since it is more or less neglected in neoclassical theory. Neoclassical discourse—if there is discourse at all—tends to be limited to prices. To interpret relationships in terms of price signalling is a bit meagre in relation to many 'real world' situations. But also, to presume that 'prices' reflect actors´ opportunity costs (marginalist theory of values) without paying attention to the institutional and power features that have a bearing on the observable prices, is an unnecessarily narrow approach to economic analysis of price formation.

Figure 3.2 should thus be understood as a general model of relationships, whereas a market relationship becomes one, admittedly important, special case. In the case of a market relationship, Actor A is replaced by Market actor A and Actor B by Market actor B in Figure 3.2. Also when referring to markets, a broad frame of reference is needed. Remember Alfred Hirschman, the economist who became well known for his rather trivial observation that in addition to 'exit' there is 'voice' as an option in communication and attempts to influence a company or other market actor (Hirshman 1970).

As suggested by Figure 3.2, actors interact, *i.e.* engage in mutual learning. They may influence each other and coordinate their activities. A relationship is normally a link in a larger set of relationships that may be referred to as a network. Each actor has a history, which will influence a specific relationship and the relationship itself has a history or development pattern from birth to maturity and perhaps decay. Many relationships relevant in relation to environmental issues have a certain continuity which sometimes can be explained by a similar ideological orientation of the two actors. The actors have their habits and this may be reflected in the relationship. Interaction and cooperative activities are largely routinized. While actors tend to look for opportunities of cooperation, conflicts of interest in relation to details or broader issues is equally a normal condition of human relationships. Dialogue is part of cooperation and can be of help in dealing with conflicts. Each actor is endowed with some knowledge and information, but this information is normally incomplete. Search efforts could add to the available stock of knowledge and information, for instance about potential relationships. To search for new relationships and to move from one relationship to another will incur transaction costs (Söderbaum 1993).

Figure 3.2 also suggests that no relationship could be seen in isolation. In addition to other actors and institutional arrangements, a physical and ecological infrastructure belongs to these surroundings. Ecosystems are here seen as a kind of infrastructure, since they deliver goods and services to humans. Building a road in the middle of a forest then no longer necessarily means that existing infrastructure is enlarged or improved, but rather that one kind of infrastructure is substituted for another.

The importance of relationships and networks has been stressed before, especially in the study of industrial markets, where illustrations that in important respects are similar to Figure 3.2 can be found. An 'interaction approach' is emphasized in a book entitled 'Understanding Business Markets. Interaction, Relationships, Networks' (Ford ed. 1990, see for instance illustration of 'relationship' on p. 20). The title of a more recent book 'Developing Relationships in Business Networks' (Håkansson and Snehota eds 1995) points in the same direction. While many of the aspects of relationships mentioned here are similar to the mentioned studies of industrial markets, there are also some important differences. The studies of industrial markets are primarily carried out at the level of companies, *i.e.* companies are in most cases seen as the actors. Individuals as actors may appear but are less visible than in the frame of reference here suggested. When reference is made to 'environment' in the studies by Håkansson and his colleagues, this refers more to market structure, position in the marketing channel, internationalization and less to the kind of 'environment' discussed in this essay. Value issues are furthermore largely avoided and concepts referring to private roles, roles as citizens, life-style or 'ideological orientation' do not appear in these studies. But many soft variables are there such as the 'image' of a corporation or its 'goodwill' in relation to customers. Inspite of the differences mentioned, the two approaches appear largely mutually supportive and compatible.

3.3   A POLITICAL ECONOMICS VIEW OF ORGANIZATIONS, MARKETS AND SOCIAL CHANGE

In neoclassical microeconomics, the principal kind of organization that is recognized is the profit maximizing firm. As part of the present approach, reference will more generally be made to a Political Economic Organization (PEO). The monetary aspect will presumably be important for all organizations, but its role in relation to non-monetary considerations may vary. An organization is a collectivity of individuals and the ideological orientation of the organization ('business concept' or 'mission statement') will depend on its socio-institutional context and the preferences and relative power of various actors and stakeholders related to the organization.

While the neoclassical firm is described as an entity controlled from a single centre, our organization is composed of individuals who differ more or less in their ideological orientations and behaviour. Business economists and organization theorists have introduced a stakeholder model of the firm or other organization, which at least in some of its versions breaks with the monistic, equilibrium thinking of the neoclassical theory of the firm. In relation to a specific organization, an attempt is made to identify all interested parties or stakeholders, *i.e.* those who have something 'at stake' in relation to the functioning and performance of the organization. Employees, shareholders, other investors, suppliers, customers, neighbours (who may suffer from pollution) are among the interest groups normally identified.

It can be noted that as part of the 'stakeholder model', each category is assumed to be relatively homogeneous. Our present model is built on Political Economic Person assumptions, where the individual is seen as an actor, and could rather be described as 'polycentric'. The organization has many centres rather than one, each individual with his or her ideological orientation representing such a potential centre. While a specific business concept may be largely accepted in an organization at a point in time, there are also some tensions between actors. And such tensions may be a precondition for creativity and the success of an organization and is not necessarily a negative thing.

As an example, some actors in the employee category may be 'Green' in orientation while others are not. Similarly some shareholders in the company may prefer a Green orientation and strategy for the company, while other shareholders are less interested. In this way Green employees, Green shareholders, Green suppliers, Green customers, Green investors have something in common and that is their ideological orientation. Similarly, actors belonging to the 'non-Green' category have something in common which represents a potential force in modifying the operations of the company. Changes in the business concept and image of an organization will often cause tensions. One network of individuals within, or otherwise related to the company in stakeholder terms, may differ in their views from those of another network, and each actor within such networks may exert his or her power to influence the course of events.

Markets can similarly be seen in the light of polycentric networks of individuals as actors. The idea is simple. For example Green consumers will prefer Green producers or companies. Green producers in turn will look for Green suppliers and so on. Market segmentation along Green lines will occur and Green networks will compete with those that are less Green or non-Green. As suggested by the Uppsala school, a relationship between one customer and one supplier could sometimes be seen as an 'organization' of its own in the case, for instance, that the two market actors enter into technological cooperation.

Reasoning in terms of ideological orientation, business concept and mission statement leads to a conditional view of the functioning of organizations and markets. Some scholars and 'practical men' have a very optimistic idea of organizations such as multinational corporations and of the market. All problems, environmental problems included, will be solved by the invisible hand. Others like David Korten (1995) warn against such dogmatic views. The 'conditional view' advocated here simply states that the functioning of organizations and markets to a large extent depends on individuals as actors in organizations and market places, their knowledge, morals and ideological orientation. Of importance is certainly also the rules of the game and other elements of the institutional context. And the rules of the game is not only a matter of laws at the national level. The actors themselves exert influence on the system of governmental rules and in addition they create their own rules. And the rules formed as part of the 'self-regulation' of business may be less permissive than the rules implemented through the state. To comply with present laws is no longer enough for many companies.

Our reasoning in terms of actors and networks also leads to criticism of neoclassical public choice theory or perhaps to a different and extended version of this theory (Söderbaum 1991, 1992). When ideological orientation is taken into consideration it becomes unrealistic to see farmers or bureaucrats as a homogeneous category. Some farmers are engaged in ecological agriculture or Green in some other sense, others do not bother much about environmental issues. Green farmers may find that they have interests in common and work together. Their interests will differ a bit from non-Green farmers. In addition, Green farmers may find that they have something in common with Green bureaucrats and may therefore engage in network building or lobbying activities with them. Predictions made on the basis of the assumptions of conventional public choice theory—for instance about a pale future for environmental interests—can be questioned along these lines.

3.4 A HOLISTIC AND IDEOLOGICALLY OPEN CONCEPTION OF ECONOMICS AND EFFICIENCY

In everyday language, and to a somewhat lesser extent in neoclassical economics, there is a tendency to associate 'economics' with 'monetary thinking'. Money and prices are at the heart of the analysis. Money is considered as a common denominator and it is assumed that it is possible to reduce impacts expressed in non-monetary terms to a monetary equivalent. No one can deny this possibility of putting a price tag on every impact. What is at issue, however, is the theoretical interpretation(s) that might be given to such monetary figure, and whether such a way of proceeding is wise or unwise under different circumstances. Wisdom in this case is as much a matter of ideology and world-view as science. My personal belief is that the 'monetary reductionism' of much neoclassical theory (*e.g.* Cost-Benefit Analysis) is part of the problems faced by the present society. The main philosophy should instead be one of holism in the sense of disaggregation and irreducibility, *i.e.* keeping different impacts separate from each other. (This emphasis does not, as we shall see, contradict the usefulness of a *partial* analysis in terms of monetary calculus or the use of prices, *e.g.* eco-taxes, as part of environmental policy.)

The use of the terms 'holism' and 'holistic' here may need further clarification. Holism can be seen as being opposed to 'atomism' and holism then refers to a broadening of the scope of an analysis. Analysis should not be limited to the micro level, for instance. Also the broader macro connections might be of interest. Analysis can furthermore be broadened from one sector of the economy to all sectors affected, using systems thinking, for example. Holism can, however, also be seen as opposed to 'reductionism'. In cybernetics and early systems theory, the reduction and understanding of biology in terms of the laws of physics was questioned. Along these lines, the reduction of complex patterns or images perceived by human beings to one-dimensional analysis can be questioned. The response suggested here is a multi-dimensional approach and analysis in terms of patterns or profiles. Something else than the one-dimensionality and 'monetary reductionism' of CBA is proposed.

Thus economics here refers to the management of all kinds of resources. Non-monetary resources (natural resources, human resources, cultural resources) are not regarded as less 'economic' than monetary ones. In other words, non-monetary resources are economic, irrespective of whether we put a price tag on them or not. And whenever we refer to the monetary price of a natural resource, only a monetary aspect is thereby taken into account. Other aspects of the same resource can be described in non-monetary (but equally 'economic') terms. It is even relevant to speak of 'resource management' in situations where no money prices are invoked at all in the attempt to arrive at wise decisions.

All economists would agree that there are monetary as well as non-monetary impacts or indicators when comparing alternative investments or development paths. As part of the reductionist view, it is proposed that non-monetary impacts and indicators can meaningfully be transformed to money numbers. This is usually done in the framework of a Cost-Benefit Analysis and the purpose is to make comparison between alternatives or between development paths simple. The monetary language is said to be accepted already in society, making the approach very practical. Those who, like the present author, are in favour of the holistic view would say that the idea of trading one impact against another in monetary terms, while simplifying things, is at the same time dangerous. At what prices should different impacts be traded against each other? What is the price of a specific irreversible negative environmental impact? At a societal level, prices and their interpretation are, in large measure a matter of politics and ideology, and the role of science should be limited to one of illuminating an issue. Rather than the technocratic role of dictating 'correct' prices for societal valuation, the scholar or analyst could choose a more democratic role. If reality is complex, why should it be treated as simple? A better strategy might be to live with some complexity.

Aggregation is less questionable at the micro level. An individual may choose his or her own weights or prices as part of an aggregation procedure. For a business company some degree of consensus about values may similarly be assumed. But even in the case of business, the disaggregative philosophy seems to be gaining ground. Monetary calculation of investment alternatives and monetary success indicators in terms of profits will be of importance as long as present institutional arrangements will remain. Institutions change gradually, however, and one example is the growing importance of a company's environmental performance. Monetary accounting and auditing will continue but environmental accounting and auditing procedures are now gaining ground. Those who are limited in their thinking to one-dimensionality will interpret non-monetary indicators exclusively in monetary terms, while others will see a value in two separate but complementary perspectives.

As part of the latter more holistic view, impacts or indicators may be categorized as in Table 3.1. In addition to the distinction between monetary and non-monetary impacts or indicators, a distinction can be made between those impacts that are expressed in terms of flows (referring to periods of time) and those that are expressed as state variables or positions (referring to points in time). This will give us four categories as indicated in Table 3.1.

The turnover of a business company and GNP exemplify monetary flows (category I), while the assets and liabilities of a corporation or the liquidity of an individual at a specific day and hour exemplify monetary positions. On the non-monetary side, pollution of cadmium from a factory per year exemplify a flow and the content of cadmium in the soil at a specific place and point in time, a position. The number of species in a forest ecosystem is another example of a non-monetary position.

Table 3.1: Four Categories of Economic Impacts (Indicators)

	Flow (referring to period of time)	*Position (referring to point in time)*
Monetary	I	II
Non-monetary	III	IV

All kinds of impacts are potentially relevant and analyses of non-monetary flows can certainly help to explain what happens to non-monetary positions. If one has to economize in measurement on the non-monetary side, measurements in terms of positions is often a good idea. A series of positions over time may be compared to make judgements about whether a lake (or the soil in some area) is approaching or departing from a 'healthy' condition. Such an interest in series of non-monetary positions or states is very different from the discounting procedure and 'present values' as part of CBA, where it is assumed that future impacts somehow can be pressed together to a point in time.

3.5 VALUATION AS BEING A MATTER OF IDEOLOGICAL ORIENTATION

For the Cost-Benefit analyst, 'value' refers exclusively to 'money value' as expressed in a set of actual prices or shadow prices. According to our premises in terms of Political Economic Person, ethics and ideology are seen as the primary basis of valuation of past, ongoing, or future activities, projects or policies. In line with this view, monetary valuation and aggregation in monetary terms become special cases which are practical and useful only under certain circumstances.

'Valuation' is then grounded in the ideological orientation of individuals in specific roles and contexts as depicted in Figure 3.3. One of our strengths as human beings is the ability to recognize patterns (Simon, 1983). Decisions can be thought of in terms of 'matching' the multidimensional 'ideological orientation' or profile of each decision maker with the likewise multidimensional 'impact profiles' of each alternative. This view opens the way for multidimensional thinking and thinking in terms of pictures and 'Gestalts', in addition to one-dimensional numbers. According to this view, an alternative is attractive for an individual if a 'good fit' between her ideological orientation and the impact profile of the alternative is expected or experienced. As part of a conscious decision process, individuals may reconsider their ideological orientation. In this sense neither the values nor the principles of valuation are seen as constant and given.

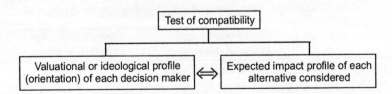

Fig. 3.3: A Holistic Idea of the Decision Act is Suggested, where the 'Ideological Profile' of each Decision Maker is Matched Against the Expected 'Impact Profile' of each Alternative.

To illustrate this idea of viewing the relationship between the decision maker(s) and each alternative considered in terms of a 'pattern recognition' or 'matching' process, an example from private life may be of help. Complex decision situations exist not only in organizations and at a public level but also, as most of us have realized, in the realm of private affairs. The members of a family may have taken a decision to move to another location and buy a new house. A number of alternatives are considered. Each family member is assumed to refer to an ideological orientation—in this case an idea of what might constitute a 'good', 'satisfactory', or even the 'best' solution to the problem faced. When visiting one of the houses considered in its specific context and when learning about its functional, aesthetic and other qualities, each family member might test the compatibility between her/his image(s) and other ideas of a good solution and the impact profile of the particular alternative at hand. If all family members experience a 'good fit', then they may be ready for a purchasing decision or continue to search in the hope of finding an even better alternative.

Whether applied at the level of private or public affairs, or something in between, the above idea of rationality does allow for pictures and images, that is visual thinking (or seeing) in addition to consideration of quantitative impacts, for instance monetary costs or physical space. There are systematic approaches to decision making other than those ending with one-dimensional numbers and experiences from private life suggests that individuals often prefer the holistic or non-reductionist idea of rationality here indicated.

3.6 POSITIONAL ANALYSIS AS A DISAGGREGATED APPROACH TO VALUATION

In Table 3.2, a way of classifying approaches to decision making and 'valuation' is suggested. Cost-Benefit Analysis is clearly highly aggregated. CBA is furthermore 'ideologically closed' in the sense that the scientist or analyst claims to provide correct rules of societal valuation (cf. category I). A highly aggregated, ideologically open version of monetary calculation (cf. Category II) might be one where each politician or other decision maker has her particular idea(s) of a good society and on this basis provides her particular price list for different kinds of impacts and makes her particular calculation of present values or internal rates of return.

Table 3.2: Approaches to Decision Making can be Classified with
Respect to Degree of Aggregation and Openness
in Ideological Terms.

	Ideologically closed	*Ideologically open*
Highly aggregated	I	II
Highly disaggregated	III	IV

The discussion in this paper points rather in the direction of a highly disaggregated, ideologically open approach to valuation of projects, programmes and policies (cf. the category IV). Positional Analysis (PA) is one such approach. This method can be characterized as follows:

- Political Economic Person assumptions and the associated actor, relationship, network approach.
- Investigation of the history and institutional context of an issue. What can be learnt from previous studies? What are the rules of the game, for instance who is making the study and according to what rules will interested parties and citizens be involved and heard? How is the present decision situation related to policy issues or other decision situations?
- Listening to interested parties and other actors with respect to problem perceptions and ideas about alternatives and relevant values or ideological orientations. Involvement of parties and actors in a search for possible solutions
- A holistic (non-reductionist) idea of economics and efficiency
- 'Systems thinking'
- 'Positional thinking'
- Conflict analysis and management
- 'Conditional conclusions' in relation to possibly relevant ideo logical orientations.

Of these characteristics, only the last four need further clarification. Systems thinking is an attempt to broaden the scope of an analysis from one sector (*e.g.* transportation) to all sectors affected. On the basis of a given set of alternatives, the analyst tries to identify those systems of various kinds that will be affected differently depending on the specific alternative chosen. The systems identified will then be helpful for the later steps of identifying differences between alternatives with respect to impacts and interests.

As already made clear-non-monetary impacts are no less important than monetary ones and it is suggested that non-monetary positions are of special significance in attempt to assess the 'healthiness' of a lake or other ecosystems or the welfare of an individual. Is the position of a natural resource improving or being degraded as a result of specific measures or projects considered? It is argued also that issues of inertia, for instance irreversibility, is best illustrated in positional terms. Actions today will influence future options.

As in the case of systems discussed above, 'activities' for individuals and organizations that will be influenced differently, depending on the alternative chosen, can be identified. For each activity identified, an assumption is made about 'goal direction' which in turn can be used for a ranking of alternatives from the point of view of the activity. For individuals living close to a road, one may assume that they prefer a level of environmental disturbances from the road (noise, pollution etc.) that is as low as possible. On the basis of this assumption and in relation to this specific activity, an alternative which implies the building of a new road, which diverts some part of the traffic, will be preferred to a situation (zero-alternative) with no new transportation facilities. A matrix can then be constructed with alternatives as columns and the identified activities with connected interests as rows, facilitating identification of conflicts of interest. It is seldom the case that one alternative is the best from the point of view of all interests identified.

Conditional conclusions are stated on the basis of matrices where alternatives are compared at the two levels of impacts and interests. Ecological sustainability (as a possible ideological orientation) may then point to one alternative as preferable whereas some other ideological standpoint may point in a different direction. As a special case, consensus may exist among decision makers about ideology. In this case, the analysis can be more closed with respect to ideology (category III in the above classification) and it becomes presumably easier to recommend one best alternative.

A number of other alternatives to CBA exist, for instance some forms of policy analysis, systems analysis, environmental and comprehensive impact assessment and multi-criteria approaches. Most versions of Environmental Impact Assessment (EIA) are 'highly disaggregated' in the sense given above, but the tradition of expertness and technocracy also often leads to aggregation in terms of points, indexes and the like. Similarly, multi-criteria approaches exist which may belong to each of the four categories according to our classification scheme. As with EIA, multi criteria approaches should be understood as a tool-box and set of methods rather than a single approach.

Among disaggregated approaches, Positional Analysis (PA), has been emphasized in the present essay. Experience of this approach dates back to the early 1970s (*e.g.* Söderbaum 1973) and are largely limited to the Scandinavian countries and languages. The method has been applied in student papers and PhD-theses (*e.g.* Leskinen 1994, Brorsson 1995) and has been developed to include control of ongoing activities and ex post valuation, so called retrospective studies (Hillring 1996). PA is one of the approaches that represent a movement away from a technocratic role for the analyst. The idea is one of involving actors and interested parties in a search for consensus while, at the same time, accepting the existence of conflicting views (cf. Dinar *et al.* eds 1995).

3.7 CONCLUDING COMMENTS

At this stage, it may be clear to the reader that even the choice of approach to valuation in itself becomes a matter of ideological orientation. Do the analyst, decision makers and interested parties share an idea of a working democracy? Do they share a view that various non-monetary impacts, such as environmental impacts and social impacts and various conflicts of interest should be made visible in the decision base or not? Are they even ready to discuss issues of power relationships, worldviews, ethics or paradigms in economics? Among ecological economists, Stephen Viederman has reminded us (1995) about Friedrich von Hayek, who in his Nobel address of 1975, "noted the irony that economists of his time were being called upon to solve the very problems that they had helped to create. Einstein noted years before that 'we cannot solve the problems that we have created with the same thinking that created them'."

Those who hold the view that public debate is important and that values cannot be dealt with in monetary reductionist terms will presumably vote for pluralism with respect to paradigms in economics and for one or other of the alternatives to CBA. It becomes especially important that the more fundamental social and environmental development issues become part of the agenda, and that a sufficient number of 'establishment actors' reconsider their world views to open up for a sustainable development path.

Such establishment actors can be found not only in business and government but also, for instance, at the universities. Will the universities be able to rearrange their strategies to play a significant role in this necessary transformation process? Will departments of economics allow a certain amount of pluralism or continue on their monistic path? Will interdisciplinary cooperation and international research cooperation in associations, such as the International Society for Ecological Economics, ISEE, contribute constructively to a sustainable development?

References

Brorsson, Kjell-Åke, 1995. *Metodutveckling av positionsanalysen genom tillämpning på Assjö kvarn. Hållbar utveckling i relation till miljö och sårbarhet* (Positional analysis applied to Assjö water mill. Sustainable development in relation to environment and vulnerability). Swedish University of Agricultural Sciences, Department of Economics, Dissertations 14. Uppsala.

Dinar, Ariel, and Edna Tusak Loehman (eds), 1995. *Water Quantity/Quality Management and Conflict Resolution. Institutions, Processes, and Economic Analyses*. Praeger, Westport.

Etzioni, Amitai, 1988. *The Moral Dimension. Toward A New Economics*. The Free Press, New York.

Ford, David (ed.), 1990. *Understanding Business Markets. Interaction, Relationships, Networks*. Academic Press, London.

Fusfeld, Daniel R. 1994. *The age of the economist*. Harper Collins, New York.

Hillring, Bengt, 1996. Forest fuel systems utilising tree sections. System evaluation and development of evaluation methodology. Swedish University of Agricultural Sciences, Faculty of Forestry, *Studia Forestalia Suecica* No 200 pp.1-17. Uppsala.

Hirschman, A.O. 1970. *Exit, Voice, and Loyalty*. Harvard University Press, Cambridge, Mass.

Howard, John A. 1963. *Marketing Management. Analysis and Planning*. Irwin, Homewood, Ill.

Håkansson, Håkan and Ivan Snehota (eds) 1995. *Developing Relationships in Business Networks*. Routledge, London.

Korten, David C. 1995. *When Corporations Rule the World*. Kumarian Press, West Hartford.

Krishnan, Rajaram, Jonathan M. Harris, and Neva R. Goodwin (eds) 1995. *A Survey of Ecological Economics*. Island Press, Washington D.C.

Leskinen, Antti, 1994. *Environmental Planning as Learning: The Principles of Negotiation, Disaggregated Decision-making method and Parallell Organization in Developing the Road Administration in Developing the Road Administration*. University of Helsinki, Department of Economics and Management, Environmental Economics Publications No. 5, Helsinki.

Myrdal, Gunnar, 1978. Institutional Economics, *Journal of Economic Issues*, 12 (4) (December) pp. 771-783.

O'Connor, Martin and Clive Spash, 1999. *Valuation and the Environment.*

Theory, Method and Practice. Edward Elgar, Cheltenham.

Sen, Amartya, 1987. *On Ethics and Economics*. Basil Blackwell, New York.

Simon, Herbert, 1945. *Administrative Behavior*. Free Press New York.

Simon, Herbert, 1983. *Reason in Human Affairs*. Basil Blackwell, London.

Söderbaum, Peter, 1973. *Positionsanalys vid beslutsfattande och planering. Ekonomisk analys på tvärvetenskaplig grund* (Positional Analysis for decision making and planning. Economic analysis on an interdisciplinary basis). Esselte Studium/Scandinavian University Books, Stockholm.

Söderbaum, Peter, 1991. Neoclassical and institutional approaches to development and the environment, *Ecological Economics*, 5, pp. 127-144.

Söderbaum, Peter, 1992. Environmental and Agricultural Issues: What is the Alternative to Public Choice Theory? pp. 24-42 in Partha Dasgupta (ed.) *Issues in Contemporary Economics. Volume 3, Policy and Development.* Macmillan, London.

Söderbaum, Peter, 1993. Values, Markets and Environmental Policy: An Actor-Network Approach, *Journal of Economic Issues*, 27 (2), pp. 387-404.

Söderbaum, Peter, 2000. *Ecological Economics. A Political Economic Approach to Environment and Development.* Earthscan, London.

Viederman, Stephen,1995. *ISEE Newsletter*, April 1995, p. 3.

4

Methods for Calculating the Costs and Benefits of Environmental Change: A Critical Review

Clive L. Spash

4.1 INTRODUCTION

Project assessment is now more commonly recognised to involve social, economic and environmental criteria, which require going beyond financial appraisal and exploring the full range of values associated with environmental change. Environmental economists take the maximum amount of money that all 'relevant' individuals might pay as the economic value of an environmental improvement, and the minimum compensation accepted as the value of an environmentally degrading action. Since the 1950s, a set of methods have been developed for producing monetary estimates of environmental change using cost-benefit analysis (CBA). The use of environmental CBA has expanded rapidly over the last three decades and applications have moved from small scale projects to entire international policies. This expansion has led to the recognition of a range of problems within the economics profession as well as criticism by non-economists.

A prominent example of the over extensions of methods is the application to the enhanced Greenhouse Effect which has also proven highly controversial (see Spash 2002). In this case monetary valuation of human life proved a particularly sensitive topic. A persistent belief is that most difficulties relate to the social benefits side of CBA because the damages avoided by pollution reduction involve issues such as valuing human and wildlife existence, cultural heritage and impacts on future generations. Costs are regarded as more straight forward because at the project level they are largely taken to involve market prices for plant construction. Cost-effectiveness is often then recommended as less problematic than CBA (Spash 1997). However, social costs as a category can cover as wide a range of entities as social benefits, and whether an item is classified as a cost or benefit depends upon the status quo and type of project. For example, the costs of building a dam include habitat destruction which may entail species loss. The regulation of flows from a dam may be undertaken to improve habitat for species and so the same species appear to benefit from such a project. If the dam is being built explicitly to avoid greenhouse gas emissions from conventional fossil fuel electricity generation then a range of secondary benefits relating to climate change damages become part of the calculation which reduces the social cost of the dam project.

There are many contentious issues arising from both the derivation and application of monetary values. An underlying concern is that the case for CBA may have been overstated. For example, among the implicit assumptions in CBA is the idea that individuals are well informed about the choices they make and their preferences are well established. Challenges to these propositions have led to an increased role for participatory approaches and the new development of deliberative monetary valuation. However, such approaches which try to maintain the status of environmental CBA create concern over the exclusion of equally valid alternatives to CBA, *e.g.* straight forward deliberation, partial ordering multi-criteria analysis etc. This paper reviews a range of methods employed for environmental CBA in order to identify some of the problems and place methods within a broader context. Throughout reference is made to commonly valued aspects of the environment which have been the subject of discussion in the valuation literature; these include: ecosystems functions, aesthetics, biodiversity, cultural and historical features and human health. These are given a brief overview in order to show the need for addressing some of the underlying features of the model of human behaviour being employed in CBA and microeconomics.

4.2 VALUES AND VALUATION

Value associated with a resource, such as a catchment, can be broadly defined in terms of economic value from direct human use (*e.g.* recreation), indirect human use (*e.g.* an option for future use, bequesting assets to future generations, maintaining the existence of species) and non-economic value which includes explicit concern over the moral standing and ethical status of entities (Spash and Carter

2002). Economic value is one of many possible ways to define and measure value in quantitative units. Although other types of value are often important, economic value is often described as providing those in the policy process (*e.g.* civil servants, government agencies) with a universal numeraire by which comparisons can be made to determine efficient resource allocation.

The basis of CBA lies in microeconomic welfare theory. Measures of economic value are based on the concept of opportunity cost. That is, when an individual exchanges a good or service they forgo an alternative exchange. How much of all other goods and services an individual is prepared to forgo in order to obtain a specific quantity of a good or service is a measure of their maximum willingness to pay (WTP). Similarly, the minimum amount of goods and services an individual accepts in compensation for losing something measures their willingness to accept (WTA). An individual's purchasing power measured in monetary units then determines their ability to command different resources affecting their allocation. This approach typically abstracts from income distribution, the allocation of property rights and availability of information all of which are crucial to the determination of actual resource supply and demand. Despite this somewhat abstract theoretical character, the environmental CBA approach has been developed as a practical tool for use in public finance.

Assessment of economic values can differ significantly from impact assessment from a natural science perspective. Under a mainstream economic approach, environmental change is linked to human welfare via characterising the environment as goods and services (which may be broadly defined). Thus, emphasis is placed upon physical impacts only to the extent that they follow a path to specific targets which affect human welfare. This means economic value categories relate to environmental goods (*e.g.* edible fish) and services (*e.g.* filtration of pollutants by reed beds, regulation of stream flow) rather than the source of physical changes (*e.g.* loss floodplain habitat, pollution). Under this approach, where environmental change takes place without affecting goods and services of use to humans there is deemed to be no economic impact.

That cause and effect may be distanced from each other by space and time and that humans are often unaware of their connection to the environment complicates the recognition of impacts and their economic valuation. The human dimension also means the expectation of an impact can be important in terms of economic welfare, even if there is no physical change, *e.g.*, 'food scares'. More generally understanding the psychological effect of environmental changes can be central to gaining insight into human well-being. In terms of preference based measures of economic welfare loss, the risk perception of individuals is all important, rather than the judgement of scientists, engineers or other experts. As long as the aim is to learn about individual's preferences and willingness to pay or accept, knowledge is required of the individual's view of the trade-off, including subjective risk assessment. Thus, for example, the fact that health and safety hazards may have been largely eliminated through tighter regulation can be irrelevant in terms of the value of the impact that the public perceives. Of course, public preferences based on mistaken beliefs or poor information may be deemed an inappropriate point of reference for policy.

Goods and services sold directly to consumers or to firms as inputs to production are traded so that standard economic models of supply and demand can be employed to estimate the economic impacts. For example, water pollution can mean the costs to the supply industry rise reflecting the opportunity cost of additional treatment of water to potable standards in accordance with governmental or international standards. Where water quality is reduced, industry reassurances may be insufficient to prevent consumer substitution away from the supplier, thereby reducing demand. In order to capture these effects a model of the water supply market would be required which allowed both consumer and producer surpluses to be estimated.

Many activities may only have indirect relationships to markets. This means aspects of an activity such as say bird watching may have some market elements (*e.g.* travel costs, entry fees, equipment costs) allowing the opportunity cost to be approximated but not the consumer surplus. The welfare in terms of surplus would require using a technique which would derive a surrogate demand function for the site's non-market values. Non-market values can be particularly strong for environmental impacts on such things as cultural heritage, biodiversity, and aesthetics. Cultural, historical and archaeological sites can be disturbed or destroyed by changes in land use. Cultural values may be associated with geological features, ecosystems (*e.g.* an ancient woodland) as well as buildings or ruins. The importance of a site in public perception can vary greatly from that of an expert, say an archaeologist or historian. Local or regional perception of a site's cultural and historical features may also diverge strongly from national or international opinion. The concept of 'landscape' and the particular characteristics of different landscapes (*e.g.*, rugged hills, gently rolling lowland valley, or river plains) can form an important aspect in people's identification with, or reaction to, a place and their perception of any changes to that status quo. The aesthetic appeal of a given environment or site involves the subjective perception of what is beautiful or stimulates the emotions. Aesthetics, or aspects thereof, have been subject to benefit estimation using HP, although the extent to which the aesthetic concept is seen to be captured will depend upon how it is defined. Providing a disaggregation of aesthetic values seems unlikely due to the difficulty of finding and agreeing upon any measure of aesthetic variation.

The functions which ecosystems perform are many and varied. Only a minority of these fall within the framework where they can be bought and sold on markets subject to private ownership. Amongst the most important functions performed by ecosystems are maintenance of climatic stability and nutrient cycles. Biodiversity of ecosystems, genes and species is seen as an important aspect of natural capital and a key to sustainable development. However valuation of biodiversity is complicated by a poorly informed general public and the extent to which people reject market valuation in this area (Spash and Hanley 1995). CBA is unable to address many of the concerns raised by the need to maintain ecosystems functions and protect biodiversity. Impacts of environmental changes in this area can be complex, highly uncertain or unknown, such as the loss of a site specific species which has never been classified.

The use of terms in the CBA literature such as total value or true value can be highly misleading because there are value categories which by definition lie outside of the economic calculus to evaluate. Economic techniques can only achieve a limited rather than comprehensive valuation of the benefits of the environment. For example, a species may be valued as a food source and because it is beautiful and because of its potential to benefit science, but it may also be valued outside and separately from all these uses or aspects of its nature which create good consequences for humans. Intrinsic values are related to non-consequentialist and therefore non-utilitarian aspects of the environment.

Economic value theory as found in standard textbooks relies upon a teleological ethical theory. Individuals who regard the world from a non-teleological perspective will express absolute values beyond any possible trade. Teleological ethical theories place the ultimate criterion of morality in some non-moral value (*e.g.* welfare, utility, happiness) that results from acts. Such theories see only instrumental value in such acts, but intrinsic value in the consequences of these acts. In contrast, deontological ethical theories attribute intrinsic value to features of the act, themselves. This could be apparent as an expression of the rights of animals to welfare or the rights of humans to life.

Intrinsic values may be regarded as rights or non-compensatory choices. Freeman (1986) has suggested that lexicographic preferences may be taken as a belief in such rights. When preferences are lexicographic, the individual cannot be compensated for the loss of a quantity of one good by increases in the quantity of one or more other goods, no matter how small the former or how large the latter (Spash 1998). However this approach reduces the difference between payment offered and compensation demanded to an anomaly within utilitarianism rather than a fundamental difference in philosophical outlook. The refusal to trade becomes particularly relevant when disruption of the environment affects such things as human health, animal welfare and ecosystems functioning and structure. In this case, intrinsic values in non-human animals, plants or ecosystems are recognised by individuals as a serious constraint on economic trade-offs (Spash 2000; Spash 2000).

4.3   METHODS FOR MONETARY VALUATION

Several methods now exist in the environmental economics literature for estimating the economic value of environmental change as enumerated in Table 4.1. Methods have moved from being market based with observed prices to creating hypothetical and politically oriented markets. The following discussion covers some of the methods which have been more commonly applied to derive monetary values for non-market environmental impacts.

4.4   THE PRODUCTION FUNCTION APPROACH (PFA)

Production functions are models that represent the physical input-output relationships involved in the production of a commercially marketed good. For instance, farmers combine environmental quality attributes such as soil fertility and water quality with purchased inputs (labour and capital) to produce crops. The physical relationships between these factor inputs and the produced output can be represented as:

$$Q = f\,(L,\ K,\ M,\ E)$$

where Q represents quantity of crop output, L and K are labour and capital inputs, M is a vector of inputs purchased in the market such as fertilizers and pesticides, and E is a vector of environmental quality variables (soil moisture and water quality, water quantity, etc.). If the marginal products of inputs ($\partial Q/\partial E$) are positive, then a decrease in environmental quality will, *ceteris paribus*, reduce output levels. Hence,

Table 4.1: Characteristics of Techniques for the Monetary Valuation of Environmental Change

Type of Methods	Market	Approach	Operational Requirements
Revealed Preference methods			
Market demand and supply	Observed	Commodities bought or sold in commercial markets.	Time series or cross sectional price and quantity data; supply/demand elasticities.
Avoided/replacement costs	Observed		
Production function analysis	Observed/Implied	Uses implicit relationships between an entity to be valued and actual expenditures on goods, factor inputs or capital assets.	As for actual markets but also detail on quantities of indirectly and non-marketed inputs. PFA uses physical dose-response functions.
Hedonic pricing	Observed/Implied		
Travel cost method	Implied		
Stated Preference Methods			
Contingent valuation method	Hypothetical	Structure questionnaires and personal interviews uses to elicit marginal WTP/WTA.	Institutional setting, realistic trade-off scenario; socio-economic and consistency information.
Choice experiments	Hypothetical		
Deliberative monetary valuation	Hypothetical	Group deliberation used to reinforce CVM or to derive group WTP for public projects.	Institutional and political setting; payment reason.
	Irrelevant		
Assumed Preferences			
Benefit transfer	Hypothetical	Take an existing value estimate and apply in different context and location.	Existing study with function which can be used to derive new estimate, but in practice often a value alone with crude adjustments to new context.

a decline in environmental quality means a loss of output for the farmer. The economic value of the change in inputs (quantity or quality) due to the change environmental quality can then be estimated by valuing the impact of observed change on the farm production function.

Specific problems which limit the use of the PFA in catchments include establishing dose-response functions and the related problem of characterising environmental quality variables. PFA generally use scientific knowledge on dose-response functions, *i.e.* the relationship between environmental quality variables and the output level of a marketed commodity. The PFA requires a quantifiable definition of the environmental quality change, linking this change to a receptor response function, and then applying the results to an economic model for a related market good. Thus, applications will depend upon the availability of existing scientific information on dose-response functions. Physical characterisation of environmental quality change requires the analysis of biological processes, technical possibilities, their interactions with producer decisions and the effect of resulting production changes on consumer and producer welfare. Biological or production response data can provide a link between water quality and the performance parameters of an ecosystem. The response relationship may be quantified directly from biological experimentation, indirectly from observed producer output and behavioural data (secondary data) or from some combination of data sources. Procedures based upon producer data (*e.g.* production or cost functions) are preferable from the viewpoint of economic analysis and can avoid the need for explicit cause-effect functions. However data and statistical difficulties have restricted their applicability and scientifically derived cause-effect functions are most commonly applied in PFA assessments. Controversies over the appropriate way in which to model responses mean that widely varying estimates of economic damages can emerge. In addition, the model must be linked to data on water quality in order to make accurate impact predictions. Such data needs to be locally or regionally disaggregated and is normally unavailable.

The PFA is unsuitable for estimating the benefits from environmental quality improvements for wildlife or recreation and tourism unless there is an associated market good or service with which to link dose-response functions. If improving environmental quality increased or reduced production costs (or increased or reduced marketed output), then a link could be feasible with the impacts on existing product supply, cost and/or profit functions. Cultural and historical values could only be assessed via physical impacts on materials which would relate only to maintenance costs rather than the socio-economic aspects of culture. The PFA is the only method which seems appropriate for addressing ecosystems functions, because of its basis in scientific knowledge which can then be linked into economic processes, although the values it will be able to assess will still be limited.

4.5 TRAVEL COST METHOD (TCM)

The Travel cost method (TCM) is the oldest of the non-market valuation techniques predominantly used in outdoor recreation modelling. An early exploration of the technique was done by Clawson and Knetsch (1966). The basic method infers the WTP to use an environmental facility from the travel expenditures of those who actually visit such sites. The travel cost demand curve is interpreted as the derived demand for the site's services (including environmental ones). For example, to evaluate recreational fishing, a TCM survey would typically gather information on travel costs, access/fish license fees, on-site expenses, and capital expenditure on fishing equipment. Varying such costs and predicting fishing activity changes can then be used to derive surrogate demand functions for fishing at a specific location.

The major strength of the TCM is that it mimics the more conventional techniques used by economists to estimate economic costs based on actual behaviour. Controversial aspects include: accounting for the opportunity cost of travel time, how to handle multi-purpose trips, and the fact that travel time can be part of the recreational experience rather than a cost. The need for user participation can limit applications. The method cannot be used to assign values to on-site environmental impacts which users find unimportant, off-site impacts or non-use values of environmental quality. There are also many statistical problems associated with the TCM. Defining the types of costs to be included in the analysis is also problematic *e.g.* setting a price per mile using marginal petrol costs or sunk costs, using the opportunity cost of time as labour price or recreational alternative. The general conclusion would seem to be that TCM researchers are forced to assign their own subjective estimation of visit costs. Randall (1994) has suggested the subjective treatment of costs means that TCM is only an ordinal measure of site value and must be calibrated using information generated from fundamentally different methods so that TCM is no longer an independent tool.

Another more general problem is the extent to which aggregating preferences reflects the type and range of values of concern. For example, the presence and size of human settlements in areas bordering a national park may play a decisive role in determining attributed monetary values. The closer an ecosystem is to large human settlements the more there are likely to be frequent visitors and hence a larger aggregate monetary value may be calculated as being associated with the site. Those coming from further away to a remote site will have a higher WTP which can counter this impact on total site value, but a green space in the city may easily prove to have a higher monetary value on the basis of low cost but frequent visits. In the extreme a catchment located in a wilderness area which restricted all access would be regarded as having no value under the TCM. Thus, contrary to a criterion of environmental prioritisation based upon the pristine or virgin status or biodiversity of an ecosystem, an altered and ecologically degraded site (accessible to local residents) can appear to be of more economic value under the TCM. Using TCM in Europe to indicate which ecosystems should be protected may therefore lead to the loss of ecosystems in remote regions, regardless of their ecological quality or significance. Graves (1991) points out that a dearth of data limits the application of TCM and HP, not only due to a lack of measurement but also because many important aesthetic features are located away from well developed markets.

The extent to which TCM can address wider aspects of environmental value is limited. Cultural and historical values only form part of a TCM estimate of demand where they are associated with site specific features, and require site visits for enjoyment. In general the site characteristics valued by TCM are only those recognised by visitors as important *i.e.*, the values are implicit in the preferences of the visitors. This means, if visitors fail to recognise the importance or even existence of a characteristic of a river (*e.g.* biodiversity) then this characteristic will be absent from the valuation via TCM. In particular, genetic diversity and ecosystem functions are unlikely to form part of site values obtained under TCM.

4.6 HEDONIC PRICING (HP)

HP is based upon the idea that any given unit within a commodity class (*e.g.* property or labour) can be described by a vector of characteristics. The basic assumption here is that prices paid for a commodity directly reflect the value of the characteristics. HP has been most commonly applied using house price sales data. In this way value aspects of say a river basin might be measured by comparing the market price of different property as affected by the characteristics of environmental quality in the river basin and a range of other factors. In essence if two properties differ only in one characteristic, the difference in prices represents the value of the characteristic. Information must be collected on a measure or index of the environmental amenity of interest, and cross-sectional and/or time series data on property values and neighbourhood characteristics for a well-defined market area that includes properties with different levels of environmental quality or different distances to an environmental amenity of interest for valuation. If data are readily available, HP can be relatively inexpensive to apply and can be adapted to consider several possible interactions between market goods and environmental quality. If data must be gathered and compiled, the costs and difficulty of application tend to increase substantially preventing application.

One of the limitations of the HP is the need for a well-informed competitive market (*e.g.* for property) within which data on environmental characteristics can be collected. Like the TCM, the method only captures people's WTP for perceived changes in environmental attributes, and their direct consequences. The method also assumes that people have the opportunity to select the combination of features they prefer, given their income. However, environmental choices are often affected by exogenous factors like taxes and government subsidies, and some endogenous factors like morality, attitudes, beliefs, and culture. The analysis may also be complicated by a number of factors such as non-linear relationships between price and property characteristics, multi-collinearity problems between variables, and choice of the functional form of the models. HP is therefore relatively complex to implement and interpret, requiring a high degree of statistical expertise and subjective judgement. The analyst must decide which factors to include as explanatory variables in the HP equation and the demand curve. Excluding a variable which has a significant effect on prices, and which is correlated with some or all of the other variables in the model, will influence the estimation of coefficients. The theoretical assumptions underlying HP mean it will give inaccurate estimates of environmental externalities if buyers lack perfect information about relevant environmental quality variables, buyers are unable to attain their utility maximising position, or the (housing) market is in disequilibrium.

HP assumes that current levels of environmental quality are the main influence on prices, but they can also be influenced by expected changes in environmental quality. For example, the prospect of improved environmental quality due to forthcoming legislation may keep property prices higher in zones near degraded environments than in the absence of such expectations. The implicit price would fail to measure the valuation of current quality levels alone.

4.7 THE CONTINGENT VALUATION METHOD (CVM)

The CVM has a great appeal among environmental economists due to its potential ability to measure a wide range of non-priced project impacts. The basic technique is to elicit bids for an identified environmental change through an interview process in a simulated market. The respondent should be familiar with the environmental change in question and accept the valuation approach. The CVM procedure can be split into six stages (Hanley and Spash 1993): setting up the hypothetical market; obtaining the bids; estimating the mean WTP and/or WTA for the sample population; estimating the bid curves; aggregating the data across time and space, and evaluating the CVM exercise. Typically, median bids are less than mean bids in CVM survey results so both should be reported although often only the mean appears. Many choices in conducting a CVM survey rely upon the judgement of the researcher including: the form of the survey (personal interview, mail survey or telephone interview); the way in which the environmental change is described; the welfare measures to be used (*i.e.*, WTP or WTA); and the 'elicitation' method by which WTP or WTA is measured (direct open-ended question, bidding game, payment cards, and dichotomous choice). A large part of the literature on the CVM is taken up by debates on several 'biases' (strategic bias, design bias, information bias) and how to interpret the WTP or WTA bids obtained from the survey. Some of the more contentious issues are described here.

Protest bids are often omitted from the aggregated bid calculation. Protest bids are zero bids given for reasons other than a zero value being placed on the environmental change in question. For example, a respondent may refuse any amount of compensation for loss of an environmental asset which they regard as unique or feel should be protected at all costs. Respondents may refuse to state a WTP/WTA amount because they reject the survey as an institutional approach to the problem, or because they have an ethical objection to the trade-off being requested *e.g.* a lexicographic preference. Initial research in this area pointed out the ethical link to rights (Spash and Hanley 1995), but has since moved on to show the prevalence of a rejection of the economic logic amongst those giving a positive WTP (Spash 1998; Spash 2000). This brings into question the meaning of the responses being obtained and how they should be interpreted.

The method of aggregating data, both across time and space, requires deciding on the relevant population, the method of aggregating from the sample bid, and the time period or discounting procedure for aggregation. The sensitivity of the results to variations in such factors should be tested and presented as a central aspect of the findings; however this is rarely the case. Discount rates as high as 25 per cent have been used without sensitivity analysis or justification (see the study made by Department of the Environment Transport and the Regions 1999).

In an hypothetical market, respondents combine information provided to them regarding the good to be valued, and how the market will work, with information they already hold on that good. Their responses may be influenced by either hypothetical market or commodity specific information given to them in the survey. This phenomenon implies that WTP/WTA values are endogenous to the valuation process. Ajzen *et al.* (1996) concluded that the information provided in CVM surveys can profoundly affect WTP estimates, and that subtle contextual cues can seriously bias these estimates under conditions where the good is of low personal relevance. However Randall (1986) has argued that CVM answers should vary under different information sets, otherwise the technique would be insensitive to significant changes in commodity framing. The divergence of opinion here relates to information provided which is regarded as important to the decision process and should therefore have an impact but fails to do so, and information which would be regarded as peripheral or irrelevant but which does have an impact on stated behaviour. Indeed, the effects of information may be inappropriately labelled as bias, depending on the way in which WTP/WTA is changed. Information which improves the knowledge of an individual concerning the characteristics of a good can be regarded as informing a consumption decision. Information which alters preferences is more problematic for the economic model which assume preferences are given. Individual preferences can hardly be regarded as exogenous to the valuation process when the goods are unfamiliar and/or never traded in a market. The idea of a neutral set of information has been shown empirically to be false with some respondents being informed while others have their preferences formed (Spash 2002). The same work shows the importance of ethical positions with regard to how information is treated.

What has been termed the embedding problem arises when the component parts of an individual's valuation are evaluated separately and when summed, found to exceed the valuation placed upon the whole. This has been attributed by some to valuation of the moral satisfaction from contributing to a worthy cause ('warm glow') rather than the good itself (Kahneman and Knetsch 1992). The counter reaction has been that CVM surveys finding embedding are flawed in some way which creates the part-whole bias, and that this can be corrected by careful survey design (Carson and Mitchell 1993; Hanemann 1994). However Bateman *et al.* (1997) have provided experimental evidence for the existence of part-whole bias for private goods outside of the CVM context. They therefore suggest the problem lies with economic preference theory rather than the CVM approach.

Many researchers have relied for guidance upon the report of the National Oceanic and Atmospheric Administration panel (NOAA 1993) and similar texts prescribing standard protocols for a good CVM study. Such guidance tends to recommend such things as personal interviews, random sampling; full information on the resource change and checks for understanding of the information given; dichotomous choice question formats; careful design and pre-testing of questionnaires, reinforcing budget constraints. Most twenty minute surveys are unable to address the full list of NOAA prescriptions and some of these lack theoretical justifications while others remain highly contentious. CVM researchers then seem to fall into two camps: those believing the method is essentially sound and needs refining for improvement, and those who see it as fundamentally flawed as a means of measuring economic values but may be useful for exploring a range of issues in economic methodology.

4.8   DELIBERATIVE MONETARY VALUATION (DMV)

The long-standing criticism of the very ontological foundation of monetary valuation, regarding individuals as consumers of environmental resources whose revealed or stated preferences determine their value, has led to a small but growing literature exploring possibilities for combining environmental valuation with deliberative (participatory) processes. Focus group discussions are argued to be a desirable component of the CVM to test survey design on the basis that group deliberation could validate the information content of the survey and help to identify other design biases. Following this logic, a new methodology, termed deliberative monetary valuation (DMV) has been advocated by Spash (2001). Niemeyer and Spash, (2001) define DMV as the use of formal deliberation concerning an environmental impact to express value in monetary terms for policy purposes, and more specifically as an input to CBA. They also provide a critical review of the theoretical foundations of DMV.

The basic idea can be explained by example, although approaches do vary. In assessing a management option the aim would be to identify all the expected socio-economic and environmental impacts and select stakeholders to meet and discuss the information. Uncertain environmental damages could be deliberated and different values and perspectives raised as in a citizens' jury. The stakeholders could then be asked to give either an individual valuation, as in a CVM survey, or an estimated average WTP of the community. The result would then be incorporated into a net present value calculation to determine the economic viability of the proposed management option.

While the added participatory approach of DMV allows increased scope for heuristic reflection over environmental values and enhances the ability to cope with complexity, drawing a boundary on the level of deliberation is required in order to draw out of the process a value. Thus, a respondent's failure to articulate a monetary valuation compatible with economic theory, while providing a point for discussion and illumination by participants (Niemeyer and Spash 2001, p. 576), leaves the economic assessment no better-off. The problem then arises that individuals may be forced to make a decision in the face of complex environmental changes. The cognitively demanding task of absorbing information about an environmental change would therefore be contrasted with requiring participants to compress all aspects into a single money metric. The expression of lexicographic preferences or non-economic reasoning is as problematic for DMV as for other stated preference methods.

4.9 BENEFIT TRANSFER

For completeness, benefit transfer is mentioned here as a technique with a rapidly growing literature. This is because of its claimed low cost and simplicity. Basically, the approach involves taking information about benefits or costs of policy changes from one context (*i.e.* source study site) and applying it to another (target) context. If this can be done, the need for primary data collection might be avoided. However, the validity of transferring numbers derived in one context to a totally different context is highly suspect (see Navrud and Bergland 2001). Basic requirements are for the original source study to have been be valid itself and for the target context to be comparable (or at least adjustable) to the source context. Good practice would dictate always adjusting the transferred cost estimates for differences in socio-economic characteristics, physical conditions of the study sites, and market conditions between the source and the target sites. A function is then required to successfully achieve this adjustment. The aim is then to transfer a cost function from the source site to the target site, or to derive a general function from analysis of a number of studies in such a way that the variations in cost estimates can be explained. In practice, monetary estimates cannot be regarded as having any meaning in terms of economic welfare. If functions are used the costs incurred in testing and deriving them can be little different than conducting an original study *e.g.* a CVM survey.

4.10 CONCLUSIONS

The qualifications and limitations of methods must be kept in mind along with the theoretical justifications for their use in the first place. HP can assess certain aspects of environmental change expost when they have been capitalised. The PFA is generally inapplicable due to a lack of scientific data. TCM is primarily concerned with recreation and tourism values at a site prior to any development. As can be seen CVM provides the most potential for comprehensive monetary valuation of environmental change and could be conducted exante or expost.

In terms of including option, existence, and bequest values, only CVM can attempt to do so. CVM also provides considerable flexibility in the types of non-market value which can be addressed in the survey. However there are several aspects of implementing CVM which restrict the extent to which it will be able to assess environmental values. In practical terms, the cost and time needed for conducting a CVM survey can be relatively high and have increased due to the extent to which various design features are now regarded as required practice. However, CVM remains an experimental technique which has been accelerated into public policy use by legal action in the USA over natural resource damages. Perhaps the greatest contribution the technique is now making is in terms of forcing economists to reconsider the content and meaning of both observed and intended human behaviour with regard to a plurality of environmental values.

In addition there are general qualifications to the use of CBA. For example, income distribution is taken as given so that prices and monetary estimates will reflect relative purchasing power in society. Adjustments could be made to the results to test for the impact of changing income distribution, but, as with other sensitivity analysis, this is rarely done in practice. The extent to which societal well-being is related to the monetary estimates obtained by the methods outlined is highly context dependent. Thus, where there are large scale changes across space and time the micro-economic assumptions underlying welfare measures are unlikely to hold. Similarly, where there are diverse cultures or differences in agreement over the role of markets there will be rejection of the value basis of the measures. Economists have tended to overlook such issues but in applied policy arenas this merely means economic prescriptions become unrealistic and academically abstract. If an ethical concern or income distribution issue is key to assessing an environmental change then a more pluralistic approach is required than offered by standard monetary valuation methods. This means explicitly considering the role of monetary estimates in an openly discussed decision process rather than assuming CBA is merely one input into an undisclosed process of politics. This has been partially recognised with the move towards DMV approaches but these seem inadequate to the task.

Existing economic valuation methods provide a series of lessons with regard to assessing environmental change. Clearly preferences can be formed on the basis of the content and the context of information provided to individuals. The motives of individuals' include a range of social, altruistic and ethical aspects which have tended to be neglected and then arise as 'problems' in valuation exercises. The standard economic specification of the policy sphere simplifies, assumes away, or makes exogenous components essential to understanding environmental problems, such as complexity, political process, and ethical considerations. Thus, CBA embodies strong implicit theories of cognition which can determine the results. In particular, ethical dimensions of social choice problems cannot simply be collapsed into desires or wants.

Generally, this review of the theory and practice of environmental valuation suggests that there is more to an individual's environmental valuation process than recognised by economic preference theory and that the value of ecosystem services cannot be restricted to economic concept of value alone. A new model needs to be tested which integrates the standard economic variables with social norms, ethical beliefs and other factors. This would offer a basis for identifying a range of drivers to explain stated preferences. In areas where an environmental change can be easily described and understood in terms of a choice based upon individual preferences, and the market trade-off implied is accepted as appropriate to the decision, then techniques such as the CVM may be applicable. Where only certain aspects of such environmental change are easily explained and/or captured in commodity terms the resulting monetary valuation will be a poor reflection of the environmental values they attempt to encapsulate and could prove highly misleading in a policy process.

References

Ajzen, I., T. C. Brown, *et al.* (1996). "Information bias in contingent valuation: Effects of personal relevance, quality of information and motivational orientation." *Journal of Environmental Economics and Management* 30(1): 43–57.

Bateman, I., A. Munro, *et al.* (1997). "Does part-whole bias exist?: An experimental investigation." *Economic Journal* 107(441): 322–332.

Carson, R. T. and R. C. Mitchell (1993). "The Issue of scope in contingent valuation studies." *American Journal of Agricultural Economics* 75(5): 1263–1267.

Clawson, M. and J. L. Knetsch (1966). *Economics of Outdoor Recreation.* Baltimore and London, The John Hopkins University Press.

Department of the Environment Transport and the Regions (1999). The Environmental Costs and Benefits of the Supply of Aggregates: Phase 2. London, Department of the Environment Transport and the Regions: 208.

Freeman, A. M. (1986). The ethical basis of the economic view of the environment. *People, Penguin and Plastic Trees: Basic Issues in Environmental Ethics.* D. van der Veer and C. Pierce. Belmont, CA, Wadsworth Publishing Company: 218–227.

Graves, P. E. (1991). Aesthetics. *Measuring the demand for Environmental Quality.* J. B. Braden and C. D. Kolstad. Amsterdam, North-Holland.

Hanemann, W. M. (1994). "Valuing the environment through contingent valuation." *Journal of Economic Perspectives* 8(4): 19–43.

Hanley, N. and C. L. Spash (1993). *Cost-Benefit Analysis and the Environment.* Aldershot, England, Edward Elgar.

Kahneman, D. and J. L. Knetsch (1992). "Valuing public goods: The purchase of moral satisfaction." *Journal of Environmental Economics and Management* 22(1): 57–70.

Navrud, S. and O. Bergland (2001). Value Transfer and Environmental Policy. *Environmental Valuation in Europe.* C. L. Spash and C. C. Carter. Cambridge, Cambridge Research for the Environment. 8: 18.

Niemeyer, S. and C. L. Spash (2001). "Environmental valuation analysis, public deliberation and their pragmatic syntheses: A critical appraisal." *Environment & Planning C: Government & Policy* 19(4): 567–586.

NOAA (1993). "Natural Resource Damage Assessment Under the Oil Pollution Act of 1990." *Federal Register* 58(10): 4601–4614.

Randall, A. (1986). The possibility of satisfactory benefit estimation with contingent markets. *Valuing Public Goods: An Assessment of the Contingent Valuation Method.* R. Cummings, D. Brookshire and W. Schulze. Totowa, New Jersey, Rowman and Allanheld: 114–122.

Randall, A. (1994). "A difficulty with the Travel Cost Method." *Land Economics* 70(1): 88–96.

Spash, C. L. (1997). Environmental management without environmental valuation? *Valuing Nature? Economics, Ethics and Environment.* J. Foster. London,

Routledge: 170–185.

Spash, C. L. (1998). Investigating individual motives for environmental action: Lexicographic preferences, beliefs and attitudes. *Ecological Sustainability and Integrity: Concepts and Approaches*. J. Lemons, L. Westra and R. Goodland. Dordrecht, The Netherlands, Kluwer Academic Publishers. **13:** 46–62.

Spash, C. L. (2000). "Ecosystems, contingent valuation and ethics: The case of wetlands recreation." *Ecological Economics* **34**(2): 195–215.

Spash, C. L. (2000). "Multiple value expression in contingent valuation: Economics and ethics." *Environmental Science & Technology* **34**(8): 1433–1438.

Spash, C. L. (2001). *Deliberative Monetary Valuation*. 5th Nordic Environmental Research Conference, University of Aarhus, Denmark.

Spash, C. L. (2002). *Greenhouse Economics: Value and Ethics*. London, Routledge.

Spash, C. L. (2002). "Informing and forming preferences in environmental valuation: Coral reef biodiversity." *Journal of Economic Psychology* **23**(5): 665–687.

Spash, C. L. and C. Carter (2002). Environmental valuation methods in rural resource management. *Nature and Agricultural Policy in the European Union: New Perspectives on Policies that Shape the European Countryside*. F. Brouwer and J. van der Straaten. Cheltenham, Edward Elgar Publishing Ltd: 88–114.

Spash, C. L. and N. Hanley (1995). "Preferences, information and biodiversity preservation." *Ecological Economics* **12**(3): 191–208.

5

Market Based Instruments for Pollution Abatement in India

M.N. Murty

5.1 INTRODUCTION

Environmental pollution is an economic externality caused by the activities related to production and consumption of goods and services in the economy. Alternatively the waste disposal services offered by the environmental media: water, air and forests could be considered as the public goods for which markets are absent. In either interpretation, the management of environmental resources could be seen as a case of market failure and therefore it is prescribed originally for the Government to intervene in the market process to control environmental externalities. Historically government intervention has taken the form of government ownership of environmental resource (example government taking the property rights over forests), use of direct regulatory measures (command and controls), and indirect measures like pollution taxes and permits (Pigou, 1920; Baumol, 1972; Dales, 1968). Experience shows that the government management of environmental externalities is a failure especially in the developing countries because of non-benevolent governments and resource constraints on meeting the high transaction cost of designing and implementing these instruments. Alternative institutions to control environmental externalities in which government plays a minimal role have developed drawing mainly from Coase's seminal contribution (Coase, 1960). Empirical experience in developing countries shows that where government regulation (formal regulation) is weak or absent, regulation by people's participation or local communities (informal regulation) has resulted in the control of environmental externalities like industrial pollution or forest degradation (Murty *et al.* 1999; World

Bank, 1999). There is also now some evidence to suggest that market agents: producers, consumers, local communities, and government have incentives to voluntarily work for the reduction of environmental externalities. This phenomenon is called a new model of pollution control (World Bank, 1999).

5.2 MARKET, GOVERNMENT AND INSTRUMENTS FOR POLLUTION CONTROL

Non-market policy instruments include command-and-controls (CAC). Market based instruments consist of pollution taxes (Pigou, 1920) and marketable pollution permits (Dales, 1968). These are often referred to as economic instruments. The choice between these instruments depends both on their efficacy in achieving the target level of emissions as well as on the relative size of welfare losses they produce (Baumol and Oates, 1988). Government can use non-market policy instruments, market based or economic instruments or a combination of two.

5.2.1 Command and Controls (CAC)

The CAC instruments are in the form of fines, penalties and threats of legal action for closure of the factories and imprisonment of the owners. They can be used either for facilitating the use of specific technologies for the environment management or for the realization of specific environmental standards. It can be shown that the cost of imposing and implementing compliance are generally higher when CAC instruments are used than with economic instruments. Furthermore, under CAC instruments, there can be no incentives for firms to innovate or invest in more efficient pollution control technologies or in cleaner process technologies.

5.2.2 Economic Instruments

Economic instruments can be divided in to three categories: price based instruments, quantity based instruments and hybrid instruments. These instruments are often called as market based instruments. Together with supply-demand forces of the market they achieve efficiency even with the presence of environmental externalities like air and water pollution.

5.2.2.1 Price Based Instruments

The price based instruments were first suggested by Pigou in 1920 in the form of taxes and subsidies to deal with detrimental and beneficial environmental externalities in production and consumption. Instances are, pollution taxes on a polluting commodity either through its production (paper, leather, electricity etc.) or consumption (cigarette, packed food etc.), or on a polluting input (fuel inputs, chemicals etc.). It could be a tax on either polluting output or pollution load. Also, there can be subsidies on the commodities the production of which generate environmental benefits (*e.g.*, neighbour's rose garden giving one the free benefit of beauty). The pollution tax or Pigouvian tax is a corrective instrument to realize the socially optimal level of economic activity generating pollution.

Pollution tax could be interpreted as the price the polluter has to pay for using the waste disposal services from the environmental media. Since the market is missing for the waste disposal service, this price could not be determined in the market. The supply and demand schedules for this service could not be observed in the market. However, given the property right to the environmental resource to the public or government, environmental regulation[1] by the government or public could make the polluter liable to pay a price for the waste disposal service. The polluter pays the price in the form of cost he incurs for complying with the environmental regulation. Therefore, the marginal cost of pollution abatement (MCA) or the cost the polluter is willing to incur for reducing every successive unit of pollution abatement could be interpreted as the demand price of waste disposal service. Figure 5.1 depicts the demand curve for the waste disposal service as the falling MCA or demand price with respect to the pollution load generated. Alternatively, it could be seen as the curve depicting the rising MCA with respect to the pollution load reduction.

There is an opportunity cost or health and other damages suffered by the public by allowing the pollution. The supply price of waste disposal service is the price charged to the polluter by the government or public for every unit disposal of waste in to the environmental media. Therefore, the marginal damages (MD) or damages from every successive unit of pollution that the public is willing to bear could be interpreted as the supply price of waste disposal service. Figure 4.1 describes the supply curve of waste disposal service as the rising marginal damages (MD) or supply price with respect to the pollution loads.

Let us illustrate the Pigouvian tax/subsidy framework diagrammatically. In Figure 5.1, MCA, MD respectively represent the marginal cost of abatement and marginal damages from pollution. E^m, E^* stand respectively for pollution loads with and without tax instrument and 't' stands for the pollution tax. With the polluters using the pollution abatement technologies, the optimality or maximization of welfare requires that the pollution be reduced up to the level at which the MCA equals the MD as shown in Figure 5.1. If a tax equivalent to 't' on per unit of pollution is evied on the polluter based on the polluter pay principle, the polluter has an incentive to reduce pollution up to the optimal level, E* in the free market. The polluter has two choices: Pay tax equivalent to E^*ERE^m or reduce pollution load from E^m to E* incurring the cost equal to E^*ESE^m If he reduces the pollution, he will save cost equal to ERS as in Figure 5.1. Therefore, given the tax rate equivalent to 't', he chooses to reduce pollution rather than paying the tax.

The damages from pollution are felt by a large number of people (more so with water and air pollution). Therefore, the damage from a unit of pollution at margin is the sum of marginal damages to all the affected people. Therefore, to design a Pigouvian tax, we require the information about abatement cost functions of polluting firms and damage functions for all the affected people. The cost of collecting the information to estimate these functions can be prohibitively high. For example millions of people are affected from the pollution of a major river like Ganges and an urban airshed like Delhi and therefore it may not be economically feasible to design the Pigouvian tax.

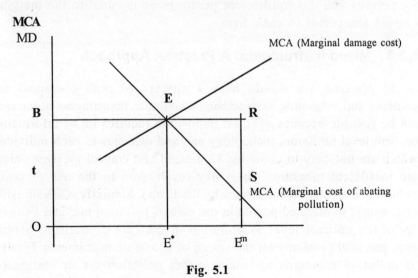

Fig. 5.1

5.2.2.2 Quantity Based Instruments

D.H Dales (1968) has suggested an alternative to the pollution tax, a system of tradable pollution rights for the management of environment.

He has proposed that the property rights be defined to the use and abuse of environment and such entitlement be offered for sale to the highest bidder. This system is like a tax to achieve the specified environment target at a minimum cost. For example in the case of air pollution, this approach first determines the optimal level pollution in a given geographical area. This level of pollution to be tolerated is then divided in to a number of permits among the various polluting units within the area (either by free distribution or by auctioning). Firms which are already comparatively more efficient in controlling their wastes or pollution (the ones that face lower unit cost for pollution control) may continue their original level of production and emissions. But they will have some extra pollution permits (or entitlements) to spare. They can sell such extra permits to firms which are less efficient in controlling their wastes (the ones that face higher unit costs for pollution abatement). Provided monitoring is possible and effective, the net result is that total pollution is kept within the prescribed levels. The more efficient firms will sell their surplus permits to less efficient firms which require more permits in order to continue with original production plans. In this process, a market for pollution permits is created in which trading in permits takes place up to the point at which the aggregate supply of permits is equal to the aggregate demand for permits and the equilibrium permit price is equal to the marginal cost of abatement to each firm.

5.2.3 Mixed Instruments: A Practical Approach

In practice, we should have a mixture of both command and controls and economic instruments. Economic instruments alone may not be feasible because of their imposition requires lot of information on firm level emission, technology etc. and damages to each individual which are not easy to come by. Command and control measures alone are inefficient measures (they may result even in the use of costly pollution abatement technologies by the firms). Similarly, the estimation of damages to affected people in the case of pollution tax, and knowing aprioi the optimal level of pollution in the case of tradable permits pose practical problems for the design of economic instruments. Fixation of pollution standards and using either pollution tax or marketable permits instrument to induce the polluting industry to meet those standards is an hybrid method using direct regulation measures and economic instruments. However, in this case the criteria for fixation of environmental standards is a subject of debate about whether they have to be decided on scientific basis or on the basis of referendum or political process.

Scientifically, they have to be based on the evidence concerning the effects of air pollution on health or of polluted water on fish and human life. They can be alternatively decided through a political process by having referendum on the choice among alternative sets of pollution standards. Still, there are issues such as should they be at state levels or national, should the standards be a compromise between the industry and people and so on.

Once the environmental standards are given apriori, the difficult problem of estimating the damages to all the affected people from pollution can be avoided for designing the economic instruments. However, we need an estimate of pollution abatement cost. It is economically feasible to obtain an estimate of pollution abatement costs because (a) the polluters may normally be much less in number than the affected people, and (b) tangible information can be obtained about technologies used by the polluters, pollution loads and levels of production. Using the firm level data on pollution loads, costs of abatement and production levels, the pollution abatement cost functions can be estimated using econometric techniques. Given the environmental standards and the estimated marginal abatement cost function, a rate of tax can be fixed such that the firms will automatically have an incentive to reduce pollution for meeting the standards. This is explained in Figure 5.2.

Let the emission standard be OE. Let the current rate of the firm's emission be OD. If the firm has to reduce pollution load from D to E as per the environmental standard, the rate of tax equivalent to OA will make the firm to do so. The rate of tax 't' in this case

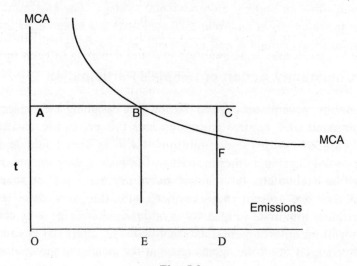

Fig. 5.2

is marginal abatement cost corresponding to the level of pollution permitted by the given standard. The firm has an incentive to do pollution abatement rather than paying tax because the cost of abatement given by the area BFDE in the figure is lower than the tax liability given by the area BCDE. Similarly marketable pollution permits can be used to obtain the reduction in pollution loads by the firms as required by the environmental standards. It can be shown that the taxes standards or tradable permits and standards method results in the adoption of least cost technologies by the firms[2].

There can be many situations in which command and control instruments are unavoidable. In several cases, the social cost of a particular activity depends on factors beyond the control of those directly involved. For example, the effects of discharge of effluents into a river depends upon the conditions of the river at that particular point of time. Similarly, stagnant air can trap air pollutants, perhaps even collecting them becomes hazardous. Therefore, exogenous meteorological conditions may contribute to occasional crisis requiring temporary emergency measures in the form of command and controls. Pollution tax rates cannot be changed on short notice to deal with emergencies and even if the changes are effected, polluters' response follow with a longer time lag. Marketable permits also result in long run adjustments in environmental quality and are not suitable for emergencies. Command and control measures on the other hand can be quickly operated to deal with more than normal amount of emissions arising out of emergencies, since they do not require extra monitoring. Therefore, in practice neither economic instruments nor command and controls alone constitute an optimal environmental strategy. The cost minimizing strategy to realize given environmental standards is a mixed one consisting of economic instruments and command and controls.

5.2.4 Community Action or People's Participation

Generally, government is given the power of designing and implementing the command and control measures and the economic instruments described above with the assumptions that it is benevolent and there are supporting legal and other instruments. In many countries, government may not be benevolent, the required environmental laws are absent and even if they are present they are ineffective due to various reasons. It is therefore important to look for institutions alternative to government for controlling environmental externalities. The alternatives can be (a) collective action of all the agents relevant for managing the environment and (b) a purely market option.

Coase (1960) has argued that many types of externalities could be optimally controlled by creating specific property rights to the resource being affected by the externality among concerned agents. Property right means right to use the resource for oneself or charge a price to others who want to use it. This important finding of Coase, now known as Coase theorem, is stated as follows: Consider a situation of an externality (say pollution). There are two agents involved here namely the generator and the affected party. Given the initial property rights to any resource either to the generator of the externality or to the affected party, and if the cost of bargaining is zero, the bargaining between the two parties results in the optimal control of externality. The final outcome of bargaining is invariant to the initial property rights (for example in the case of air pollution whether the right to clean air is vested in the affected people or whether the right to pollute is given to the polluter). This result is further explained in Figure 5.3. In this figure, pollution load is measured along x-axis and the marginal cost of pollution abatement (MCA) and the marginal damages (MD) are measured along y-axis. The optimal pollution load is given as OE. For the pollution loads higher and lower than OE, there are incentives for gainful bargaining between the polluter and the affected party. If the polluter has the right to pollute beyond OE, then the MD is higher than MCA for the pollution loads, the affected party has an incentive to bribe the polluter at any rate lower than MD for a unit reduction in pollution and the polluter has an incentive to accept the bribe at any rate higher than the MCA. Therefore, bargaining between the two parties takes place until the pollution load is reduced to OE. Similarly, since MCA is higher than MD for the pollution loads lower than OE, the polluter has an incentive to offer bribe to the affected party at any rate lower than MCA and the affected party has an incentive to accept bribe at any rate higher than MD. Again, the bargaining between them leads to the optimal pollution load OE.

There are several practical problems for the Cosean bargaining to work in practice for controlling the environmental externalities. First of all, in reality, the transaction costs or costs of bargaining are not zero but positive. It can be shown that with positive costs of bargaining, the resulting pollution load through bargaining can be higher or lower than the optimal pollution load 'OE' depending on the initial property rights.

That means with the positive transaction costs, the final result will be no longer invariant to the initial property rights. Secondly, one of the key assumptions in the Coasean solution is that all the externalities

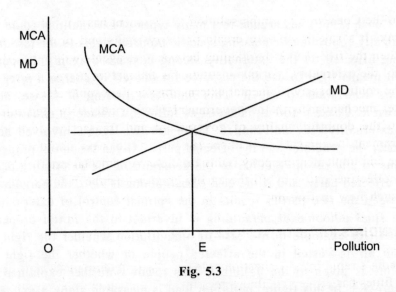

Fig. 5.3

are captured in the value of property rights and there are incentives for gainful bargaining. This can work well for the externalities on a smaller scale or local externalities of the type described by Coase (a building that blocks wind mill's air currents; a confectioner's machine that disturbs doctor's quiet etc.). However, many environmental externalities occur on a grander scale with a large number of receivers and many times a good number of generators (for example pollution of a river and the atmosphere) which makes defining property rights and facilitating bargaining difficult. One way of dealing with this problem is to create a common property right to the river for all the affected people as one group and have an association of polluters of the river so that the bargaining to reduce the river pollution can take place between the two parties. The third problem for Coasean bargaining arises again in the context of defining the property rights for an environmental resource. The environmental resource is a stock affecting the welfare of both present and future generations. Capitalization of future benefits from this resource is not possible because property rights to future generations of affected people can not be defined. One approach to take care of future generation is to consider the government as its representative. The government can compete in the market for environmental property rights of the future generation and pay for it by issuing a debt which has to be serviced by the future generation. Another approach is based on the assumption that the present generation has a bequest motive to the future and wants to bequeath to the future the preserved resources. However, both government intervention and bequest motive are out side the scope of Coases's property rights approach.

In the Coasean bargaining solution, government has a minimal role to play. It's role is only to create property rights and protect them and then the free market bargaining between the agents will optimally control the externality. Various institutional alternatives now considered for the control of environmental externalities contain some elements of market mechanism with the government playing only a limited role. Given the doubtful quality of government and transaction costs of government instruments, it is imperative to look for new institutions to define and implement property rights for the environmental externalities. The collective action by all the agents involved has been found to be one such new institution.

5.3 INDIA'S INDUSTRIAL POLLUTION

India's rapid industrialization aided by the economic planning since mid- fifties has created enormous pressure on its environmental resources. The environmental resources like water, land, and air have become dumping grounds of industrial waste with toxic materials in the absence of environmental law protecting them. Another environmental resource, the forest, has been subjected to rapid degradation due to growing demand for forest based inputs from the industrial development.

Rapidly growing demand of the industrial products for the intermediary and final uses in the post independent India has contributed to the growth of a number of industries which are more intensive in the use of environmental resources. There has been a phenomenal growth of industries like Chemicals, Fertilizers, Paper and Pulp, Sugar, Distilleries, Oil Refineries, Coal Fired Thermal Power Generation among others during the last forty years.

First half of almost fifty of years of development planning in India while contributing to rapid industrialization, especially through the promotion of heavy industry, was silent about the sustainable use of environmental resources. By the time the Government of India has realized the need for environmental regulation as an integral part of it's strategy for rapid industrialization, considerable damage was done to India's valuable natural resources like forests, fertile lands and water bodies. The planning period constituting two decades during 1970-1990 has seen a spate of legislations (many of which were later amended) providing for the environmental regulation in India. These legislations have empowered the Central and State Governments to create necessary institutions and instruments and monitor their functioning. In spite of more than thirty years of environmental legislation, India is lagging behind in the use of instruments like pollution taxes, the institutional or technological options like promoting collective action and the implementation of environmental law protecting the public interest.

Ever since economic planning started in India attempts have been made to integrate the management of environmental resources with the National Economic Planning. Some resources are earmarked for environmental protection out of plan funds for many development projects, especially projects for energy and other infrastructural development. Certain measures are taken for the environmental safety and the promotion of environmental friendly technologies in the industrial production. Industries have been asked to prepare a detailed report identifying the sources of pollution by activity and measures taken to control it. All projects above a certain size and located in environmentally fragile areas have been asked to obtain the clearance from the Ministry of Environment and Forests.

5.4 INSTITUTIONS FOR INDUSTRIAL POLLUTION ABATEMENT IN INDIA

Government of India has started showing some concerns for its natural environments only since early seventies. The enactment of environmental laws, Wildlife Protection Act, 1972 and Water (Prevention and Control of Pollution) Act, 1974 has marked the beginning of an eara of governmental regulation for the control of environmental pollution in India. In continuation, Government of India has undertaken a number of regulatory measures including the Water (Prevention and Control of Pollution) Cess Act, 1977; the Forest (Conservation) Act, 1980; The Air (Prevention and Control of Pollution) Act, 1981; the Environment (Protection) Act, 1986; the Motor Vehicle Act, 1938, amended in 1988 and the Public Liability Insurance Act, 1991.

Out of all the legislations mentioned above, the Water (Prevention and Control of Pollution) Act, 1974, amended in 1986; the Water (Prevention and Control of Pollution) Cess Act, 1977, amended in 1988; the Air (Prevention and Control of Pollution) Act, 1981, amended in 1988 and the Environmental Protection Act, 1986 are the most important legislations pertaining to industrial pollution abatement in India. The Water Act has resulted in the creation of Central and State Pollution Control Boards with the aim of prevention, abatement and control of water pollution. The Pollution Control Board can demand information from any person or industry to guarantee the compliance with the Act. This information could include the discharge of trade effluent in to a stream or on land and details about the installation and operation of pollution control equipment. Non-compliance with the directions of the Boards is punishable by imprisonment up to three months and a fine of up to Rs. 1000. An additional daily fine of Rs. 5000 can be imposed for the continued non-compliance. If the non-compliance continues beyond a period of one year after the date of conviction, the offender can be punishable with imprisonment for two to seven years with fines. The Board, if necessary, can also close down certain factories that create pollution.

Air Act, 1981 with it's amendments in 1988 empowers the Central and State Boards to prevent, control and abate air pollution. The principal functions of Central Board are to advise the Central Government on the matters pertaining to improvement of air quality and the prevention and control of air pollution, to coordinate the activities of State Boards and provide technical assistance to them and to lay down national standards for air quality and collect and publish data related to air pollution. State Boards have similar functions at the State level. State Government on the advice of State Board may declare any area as the air pollution control area and prohibit the use of any fuel that may cause air pollution in that area. Also, State Government should ensure the standards for emissions of air pollution from automobiles laid down by the State Board. No person shall, without the consent of State Board construct any air polluting industrial plant in the pollution control area. Failure to comply with the provisions of this Act by any individual can make him liable to imprisonment up to three months or fine up to ten thousand rupees or both. Continued non-compliance after the conviction can make the individual liable to a fine of hundred rupees per day.

Water Cess Act, 1977 with it's amendments in 1988 provides for levying a cess on water consumed by certain industries and by local authorities. The main objective of this cess is to increase the resources of Central and State Pollution Control Boards for the prevention and control of water pollution. The rate of cess applicable to various water polluting activities vary from Rs. 0.015 to Rs. 0.08 per kilo liter of water consumed. If some or all the activities liable to the payment of water cess install effluent treatment plants, such activities are entitled to a rebate of seventy per cent of the cess payable. The proceeds of the cess will be first credited to the Consolidated Fund of India and the Central Government may pay to the Central Board or State Boards after deducting the cost of collection. The Central Government has to take into account the cess collected by the State Government in deciding the amount payable to the State Board. Willful evasion of cess by the concerned parties can make them liable to the imprisonment up to six months or fine up to thousand rupees or both.

Environment Protection Act, 1986 is intended to remedy the lacunae noticed in the earlier laws and to serve as a single environmental legislation. This Act provides for the protection and improvement of environmental resources like water, land and air as per the decision taken at the United Nations Conference on the Human Environment held at Stockholm in June, 1972. This Act empowers the Central Government to take all such measures as it deems necessary for the protection and improvement of the quality of environment. In particular, the powers of Central Government include the co-ordination of actions taken by the State Governments, planning and execution of nation-wide programme for the pollution abatement and laying down standards for emission of environmental pollutants from various sources.

5.5 FISCAL AND OTHER INCENTIVES FOR THE CONTROL OF INDUSTRIAL POLLUTION IN INDIA

Threats of closure of a polluting activity and penalties and imprisonment for the offending parties as envisaged in various legislations described above are the command and control type of regulatory measures. Even though it is widely known that command and control measures do not provide the necessary incentives to the polluters for the choice of least cost methods of pollution control (Cropper and Oates, 1992), the Government of India has so far resorted only to such measures for controlling industrial pollution in India. On the other hand, the fiscal instruments like pollution taxes or marketable pollution permits which are used in some developed countries, though also coercive, provide incentives to the factories for adopting the least cost pollution abatement technologies. Ironically, there are no serious attempts made in India so far for using such instruments for industrial pollution abatement. The currently levied cess on the consumption of water by the industrial activities could not be treated as a pollution tax, since, as mentioned above, it's main objective is to raise revenue to the Pollution Control Boards. As such, the cess collected is very nominal (Rs. 0.015 to 0.07 per Kl) which normally does not have much effect on the demand for water by the industry. Some of the recent research studies on the water pollution abatement in India (Dasgupta and Murty, 1985; Gupta, Murty and Pandey, 1989; Mehata *et al.* 1995; Murty *et al.* 1999; Murty and Surender Kumar, 2004) have found that the pollution tax on the industrial water use should be several times higher than the current rate of water cess to realize the prescribed water quality standards.

Another anomaly of the choice of instruments for the industrial pollution control in India is the widespread use of subsidies for the polluting parties. It is well known that pollution taxes have to be imposed on the external diseconomy generating or polluting activity to make a necessary correction for the market failure. On the other hand, pollution subsidies provide perverse incentives to them for increasing their scale of production and hence the environmental pollution (Baumol and Oates, 1979). Government of India has been providing pollution subsidies to the industries in various forms. The pollution control equipment has been given the benefit of accelerated depreciation for corporate tax computation. There has been also an investment allowance for the corporate tax calculation for the investments on the pollution control equipment. There is another direct tax concession for the pollution control in terms of exemption of capital gains tax for the capital gains obtained through the relocation of a polluting factory. There are also rebates of excise and customs duties for many types of pollution control equipment. These tax concessions or subsidies are not justified either on efficiency or equity grounds. They do not guarantee the allocational efficiency because they do not provide the polluter, unlike pollution taxes, the incentive to choose the lease cost pollution abatement technologies. They have the effect of increasing the polluters' profits which is not desirable in a developing country since the big factory owners/ polluters normally belong to higher income groups.

There is one important case in which subsidies or financial incentives for the industrial units are justified for the control of industrial pollution in India. During early sixties, several State Governments in India have come out with the legislations in aid of industrial estates with the objectives of promoting small scale industries and achieving the balanced regional development. More than 800 items are reserved for the production by the small scale enterprises. The small scale enterprises are also entitled for the financial assistance, tax benefits, and subsidized electricity and water charges. As a result of this policy of Government, the industrial estates in various parts of the country are dominated by small and medium scale industries creating a number of environmental problems. Some of these problems are: inadequate understanding of the technology of waste generation and treatment, lack of required space for the pollution control facilities, ineffective supervision and management of even simple installations for the pollution control and the lack of technical assistance. Therefore, governmental assistance to small scale enterprises may be justified for helping them to circumvent these problems.

5.6 COLLECTIVE ACTION IN INDUSTRIAL WATER POLLUTION ABATEMENT

Presence of scale economies in the pollution abatement especially in the water pollution abatement, have compounded problems for the industrial estates in India. In such a situation, first of all, it is not economical for the small scale enterprises to have their own effluent treatment plants. Secondly, pollution taxes are ineffective because the factory chooses to pay tax (tax payment will be lower than the cost of treatment in this case) rather than spending on pollution abatement.

Collective action involving all the relevant parties namely factories, affected parties and government is now seen as an institutional alternative to deal with the problem of water pollution abatement in industrial estates in India (Murty *et al.* 1999; World Bank, 1999). Collective action in industrial water pollution abatement is meant for bringing necessary institutional changes that are compatible with the choice of cost saving technologies in comparison to the technologies chosen in the free market. For example, the technology of Common Effluent Treatment Plant (CETP) for an industrial estate can be adopted if necessary legislations are made to define the property rights of the factories and the affected parties and the active political groups of clubs of factories and the associations of affected parties are formed. The emergence of active political groups depends upon the benefits each group gets from the bargaining. A CETP for an industrial estate confers the benefits of saving in costs to the factories and the reduction in the damages to the affected parties.

Collective action in the industrial pollution abatement may not be possible due to various factors that constrain the relevant agents' willingness and ability to participate (Murty *et al.* 1999). Some of these constraints can be in the form of

(1) Inadequate and ambiguous environmental law protecting the rights of pollutees to clean environment.

(2) Lack of public awareness about the magnitude of damages to the community from the environmental law protecting the rights of the people.

(3) Lack of resources to local communities to organize themselves as politically active groups for taking recourse to legal action against the polluters.

(4) Inadequate resources at the disposal of small enterprises to organize themselves as a club for setting up a CETP.

(5) The problem of agreeing upon a cost sharing arrangement for a CETP by a club of factories.

(6) The problem of designing the methods of charging the factories and households for treating the effluents by a CETP.

(7) Political uncertainty in the case of clubs of pollutees and the uncertainty of continuance of agreement for cost sharing by a club of factories.

Government can play an enabling or catalytic role in the collective action by aiming to remove some or all of these constraints. In a democracy, the legislatures can enact environmental laws and create legal institutions to protect the rights of the citizens to clean environment. Public awareness about the magnitude of damages from the pollution can be promoted by creating the institutions for environmental education. The Government of NGOs can even provide subsidized legal facility to the local communities. It can give small enterprises concessional financial loans and technical help for setting up CETPs. NGOs can provide expertise to a club of factories to work out mutually agreeable methods of sharing the cost of CETP.

There are many incentives for polluters, affected parties and the Government to promote collective action in the industrial water pollution abatement.

Incentives for the industries:

(1) CETP is less expensive in terms of capital and operation cost.

(2) It is easy for a club to secure financial support from Government and NGOs.

(3) With CETP, water can be treated economically to produce process grade water which can be reused and recycled in the industry. It is an important incentive to form a club in the water scarce regions.

(4) The size of an industrial estate depends upon the water availability and the facility to dispose the waste water on a sustainable basis. The CETP may help achieve both these objectives.

Incentives to the affected parties

(1) Improved quality of drinking water.

(2) Reduction in the damages from water borne diseases.

(3) Recreational facilities from the preserved water body.

(4) Reduction in the cost of legal action against polluters.

(5) Increase in the access to legal institutions.

Incentives to government

(1) Reduction in the burden of various Government agencies working for abating and controlling the water pollution.

(2) Respectability for the catalytic role rather than the unpopularity for the coercive role.

The observed complementarity between the collective action in the industrial water pollution abatement and the interests of all the relevant parties indicates that the collective action involving the factories, affected parties and the Government is feasible. It also highlights the roles of different parties, especially the catalytic role of Government. Some recent studies (Murty, 1994) on the collective action possibilities for the management of environmental resources have shown that similar type of complementary relationship exists between the collective action for the preservation of local commons and forestlands and the interests of local communities and the Government. As another example, collective action of national governments is explained as an institutional arrangement for designing and implementing the international carbon tax to contain global environmental pollution (Hoel, 1991; 1992; Murty, 1995).

Collective action as described above is now emerging as an institutional alternative to deal with the problem of water pollution abatement in a large number of industrial estates in India. Recent studies (Murty *et al.* 1999) in India about the historical developments leading to the adoption to CETP technologies by some industrial estates in Andhra Pradesh, Haryana and Tamil Nadu states clearly provide evidence for the role of collective action involving affected people from pollution, the factories, NGOs and Government. These studies show that given the environmental laws protecting the interests of affected people and the factories, the collective action of affected people precedes the collective action of factories. The collective action of affected people acts as a deterrent to factories compelling them to undertake pollution abatement. Given this deterrent, the collective action of factories is necessary to adopt the cost saving CETP technology.

The empirical experience in some developing countries shows that the collective action or the environmental regulation by the local communities (informal regulation) is emerging as an alternative to the regulation by the government (formal regulation) even in the case of controlling pollution by the isolated big factories (Murty *et al.* 1999; World Bank, 1999). A new model of pollution control explains that market itself provides incentives for various agents to voluntarily take interest in reducing industrial pollution (World Bank, 1999). The producers have a stake in improving the environmental performance of their industry because some empirical studies in developed countries have shown that equity share prices of industry are positively related to its environmental performance. Consumers by voluntarily abstaining from buying the products of a polluting industry could make the industry to improve it environmental performance. The local communities could bring pressure on the industry through collective action. The Government could play an enabling role in promoting R & D in the pollution abatement technologies and providing legal aid and environmental education for collective action by the local communities.

5.7 FEASIBILITY OF POLLUTION TAXES IN INDIA

In the context of tax reforms leading to the Value Added Taxation in India, it is important to examine what can be the form and shape of taxes meant for the abatement of environmental pollution. It is especially important in the case of reforms leading to the free market production and supply of most of the commodities and services in which fiscal instruments have to play an important role in controlling pollution. The taxes for pollution abatement can be of three types: (a) a tax on the commodity, the production of which generates pollution, (b) taxes on the inputs, the use of which contribute to the pollution and (c) a tax on the effluents. Since commodities differ with respect to the pollution intensity in their production and use, the pollution taxes should be commodity specific leaving the tax structure more differentiated even after having a general revenue VAT with a uniform rate. Even a tax on a kiloliter (KL) of residual water in the industry to realize the given environmental standards cannot be uniform across the industries since the pollution loads in a KL of water differ across the industries. In contrast, pollution taxes on the effluents may end up with less differentiated tax structure than the pollution taxes on the commodities. Taxes on the effluents say, tax per unit of BOD, or tax per unit of COD can be common for all the industries given the environmental standards.

Optimal pollution taxes on the effluents can be uniform across the industries and the regions (states in a federation) if all the industries have access to the available pollution abatement technologies and the effluent standards are common for all the industries (states). In actual practice, the ambient environmental quality may demand different standards for the different regions. Suppose the ambient water quality in Andhra Pradesh may be such that it can have less stringent water quality standards at source than Maharastra or Gujarat. Then the rates of effluent taxes required to realize the effluent standards will be lower in Andhra Pradesh than in the other two states. If the industry can shift the entire cost of water pollution abatement to the consumers of it's products, the prices of water intensive commodities in Andhra Pradesh will be lower than the prices of their counter parts in the other states. In other words, ceteris paribus, Andhra Pradesh enjoys the comparative cost advantage in the production of water intensive commodities like paper and pulp, fertilizers, tanneries, distilleries etc. However, if there are no cross border restrictions on investment and trade, market competition can locate the industries such that per unit pollution abatement costs are equalized across the regions. The unrestricted cross border trade and investment equalizes the ambient environmental quality across the states. For instance, the water abundant state attracts the water intensive industries up to the point where it's water quality is deteriorated to the level of water quality of the other states so that it no longer enjoys the comparative cost advantage in the production of water intensive commodities. The equalization of ambient environmental quality across the states results in the common effluent standards and taxes.

Pollution abatement is a production activity preserving the environmental resources which in turn provide a number of benefits to the society on a continuous basis. A preserved environmental resource like a lake is a public good which may be treated as a primary input in the waste disposal activities and the activities supplying the environmental amenities. To identify the value added benefits of this primary input, the lake as a public good, let us consider the situations in the economy with and without the lake. Also, assume that in both the situations, a factory draws water from other sources for use in it's production processes and uses the lake only for waste disposal if it is there. In this case, the value added by the lake is savings in the waste disposal cost which may be incurrred by the factory if lake is not there. Since the lake, as an environmental resource, is a stock of natural capital, the savings in the pollution abatement cost to the factory due to the lake may be treated as what is the value of price of capital services for the stock of man made capital. However, unlike man made capital, the environmental resource is not owned by the factory but it belongs to the society at large. Then, the important question is whether one can treat the value added by lake as part of private profits of the factory or it has to be collected by the Government from the factory in the form of a cess or royalty for the waste disposal services rendered by the lake. Also, if the ambient water quality standard of the lake falls short of the prescribed pollution standards so that the factory has to incur certain abatement cost, the value added by the lake is given as the difference between the current abatement cost and the cost of next best alternative abatement method. In this case, the payment of the factory for the environmental services consists of royalty plus a pollution tax or the marginal cost of abatement. However, in a free market situation, any factory can use a given water body for it's waste disposal without paying for it. Only in the presence of fiscal instruments like pollution taxes or any other institutional arrangements, the factory is made to pay for the services of the water body as a waste receptor. Of course, the total value added by a public good like a preserved lake will be higher than it's value addition in the industry alone since there can be value additions in fisheries and in the activities promoting the water based environmental amenities.

5.8 OBJECTIVES AND STRATEGIES OF THE ENVIRONMENTAL POLICY IN INDIA

Evolution of environmental policy and institutions to protect the environmental resources has been providing important inputs to integrate the management of environment with the national economic planning in India. The first formal recognition of the need for the integrated environmental planning was made when the Government of India constituted the National Committee on Environmental Planning and Co-ordination (NCEPC) in 1972. Till 1980, the NCEPC has done valuable work in a number of areas like environmental appraisal of the projects from selected sectors and spread of environmental awareness. During the Sixth FiveYear Plan in India (1980-85), there was an explicit recognition that environmental factors and ecological imperatives have to be built in to the total planning process to achieve the long terms growth of sustainable development. The major areas in the environment management in which work was initiated during the Sixth Plan include water and air pollution monitoring and control; environmental impact assessment; environmental research promotion and environmental information, education, training and awareness. Significant progress was made in the management of environment and ecology during the Seventh Plan period (1985-90). The Pollution Control Boards (Central and State) undertook major tasks of controlling pollution at source, development and operation of national air and water quality network and the implementation of national pollution standards. Programmes on waste recycling and prevention of coastal pollution were also undertaken.

Efforts of Government of India for protecting environmental resources on the backdrop of planning for rapid industrialization in India have been further intensified in the Eighth Five Year Plan (1992-97). Some of the major tasks for the environmental planning as indicated by the Eighth Five Year Plan include among other things the decentralization of control over nature and natural resources, making individuals and institutions more accountable to the people for their actions impinging on environment and ecosystem etc. The strategies to achieve these objectives should be comprehensive so that they are appropriate for different geographical regions and ecological and social systems. There can be preventive and regulatory strategies. The preventive strategies consists of raising public awareness, strict enforcement of laws, statutory efforts to regenerate the productivity of the ecosystems. In contrast, the strategies of regulation are applicable to expose situations where environmental damage causing activities have already started. The strategies for regulation include among other things public participation and involvement of NGOs in prevention and control of pollution, prevention of environmental degradation by providing necessary technical help through designated institutions, setting up of appropriate machinery by Central, State and Local Governments for speedy response to investigation and disposal of public complaints etc.

As expected, there is a complementarity between the objectives and strategies for the environmental management stated in the Eighth Five Year Plan document and the action points for integrating environmental considerations with the strategies for industrial development given in the National Conservation Strategy and Policy Statement on Environment and Development, Government of India, Ministry of Environment and Forests, 1992. The action points given in this statement are as follows:

(a) Incentives for environmentally clean technologies, recycling and reuse of wastes and conservation of natural resources;

(b) Operationalization of 'polluter pays principle' by introducing effluent tax, resources cess for industry and implementation of standards based on resource consumption and production capacity;

(c) Fiscal incentives to small–scale industries for pollution control and for reduction of wastes;

(d) While deciding upon sites, priority to compatible industries so that, to the extent possible, waste from one could be used as raw material for the other and thus the net pollution load is minimized;

(d) Location of industries as per environmental guidelines for setting up of industry;

(e) Enforcement of pollution control norms in various types of industrial units depending on their production process/ technologies and pollution potential; particular attention to be paid to highly polluting industries;

(f) Encouragement for use of environmentally benign automobiles/ motor vehicles and reduction of auto-emissions;

(g) Collective efforts for installation and operation of common effluent treatment facilities in industrial estates and in areas with a cluster of industries;

(h) Introduction of environmental audit and reports thereof to focus on environment related policies, operations and activities in industrial concerns with specific reference to pollution control and waste management;

(i) Dissemination of information for public awareness on environmental safety aspects and stringent measures to ensure safety of workers and general pollution against hazardous substance and process;

(j) Preparation of on- site emergency plans for hazardous industries and off-site emergency plans for districts in which hazardous units are located;

(k) Public liability insurance against loss or injury to life or property;

(l) Setting up of environment cells in industries for implementing environmental management plans and for compliance of the requisites of environmental laws;

(m) Internalizing the environmental safeguards as integral component of the total projects costs;

(n) Environmental impact assessment from the planning stage and selection of sites for location of industries; and,

(o) Clearance by Ministry of Environment and Forest of all projects above a certain size and in certain fragile areas.

5.9 CONCLUSION

Assimilative capacity of environmental media and the damages to the public from pollution determine the supply of waste disposal services while the demand for them is determined by the cost of pollution abatement to the industries given the environmental regulation. Pollution tax is nothing but the price that equates the supply and demand for environmental services. Given the pollution standards based on the assimilative capacity of environmental media, the supply of waste disposal services is fixed. The pollution tax or price is then determined by the demand price or the marginal cost of abatement for the waste disposal services. This method of designing the pollution tax is called the tax-standard method.

Management and preservation of environmental resources did not form part of strategies for economic development of Government of India at least up to early seventies in the post independent India. Only the later half of the planning period has witnessed the growing concern of the policy makers and the general public about the state of environmental resources.

The process of integrating the management of environmental resources with the National Economic Planning has started with the Sixth Five Year Plan. The subsequent five year plans have seen the issues of preservation of environmental resources and the sustainable economic development in India as important an objective as many other development objectives.

Two decades constituting mainly seventies and eighties have witnessed a spate of governmental legislations creating the environmental laws protecting the environmental resources and the interests of general public, and making the polluters liable to the damages to the environment. These legislations have contributed to the creation of Central and State Pollution Control Boards to prevent and monitor the environmental pollution.

In spite of having the necessary law and institutions to protect the environment, ironically India is lagging behind in the design and enforcement of economic instruments like pollution taxes. The water cess on the industrial use of water that is currently levied by the Government is nominal. Some Recently carried out research studies on water pollution abatement in India have shown that the rate of pollution tax on the use of water to realize the prescribed standards can be much higher than the currently levied water cess.

Also for the sake of designing effective environmental pollution control policy in India, it is important to realize that there are cases in which pollution taxes are not effective. There has been a phenomenal growth of industrial estates containing small and medium size enterprises during the last two decades in India. Due to presence of economies of scale in the pollution abatement, it is not economical for the small factories to have their own effluent treatment plants and the pollution tax can not induce them to spend some maney on pollution abatement. In such a situation, collective action through the clubs of polluting factories or the clubs of affected parties / municipal committees have to be encouraged for making use of the technology of common effluent treatment plant. Rightly, Government of India has taken up certain steps to encourage the setting up of common effluent treatment plants for the industrial estates in India.

The policy of Government of India to give subsidies in various forms to big factories to encourage them to have effluent treatment plants runs against the basic economic tenet of taxing the external diseconomy creating activities for the efficient allocation of resources. Subsidies to polluting factories will provide them perverse incentives to expand their scale of operation and there by pollution. However, subsidies may be justified to small enterprises from the point of view of equity and efficiency. Since the pollution tax on a small enterprise is not effective, a subsidy or a financial incentive to small enterprises to encourage the use of technology of the common effluent treatment plant can be efficient from the point of view of resource allocation.

In the context of reforms leading to value added taxation in India, pollution taxes based on per unit of a commodity or per Kl of residual water are more differentiated so that they may not go well with a VAT having one or two rates. However, taxes on effluents say, tax per unit of BOD or tax per unit of COD can be common for all the industries given the standards.

A factory may have to pay royalty over and above the pollution tax for using a water resource for the waste disposal from the point of view of equity if the society has the property right to the water resource but not the factory.

End Notes

1. In the current literature, the regulation by the government and the public are respectively regarded as the formal and informal regulation. See World Bank (1999).
2. For more details, bee Baumol and Oates, 1988.

References

Baumol, W.J. (1972): "On Taxation and Control of Externalities", *American Economic Review*. 42 pp. 307-322.
Baumol, W.J., and W. Oates (1989): *Economics, Environmental Policy and the Quality of Life*, Englewood Cliffs, N.J. Printice Hall.
Central Pollution Control Board (1989): *Pollution Control Acts & Rules with Amendments*, New Delhi.

Coase, R. H. (1960): "The Problem of Social Cost", *The Journal of Law and Economics.* 3, pp. 1-44.

Cropper, M.L., and W.E. Oates (1992): Environmental Economics: A Survey, *Journal of Economic Literature*, Vol. XXX, pp. 675-740.

Dales, J.H. (1968): *Pollution, Property and Prices.* University of Toronto Press, Toronto.

Dasgupta, A.K., and M.N. Murty (1985): "Economic Evaluation of Water Pollution Abatement: A Case Study of Paper and Pulp Industry in India", *Indian Economic Review*, Vol. 20, No.2, pp. 231-267.

Gupta, D.B., M.N. Murty and R. Pandey, (1989): *"Water Conservation and Pollution Abatement in Indian Industry: A Study of Water Tariff"*; National Institute of Public Finance and Policy, New Delhi.

Hoel, M., (1991): "Global Environmental Problems: The Effects of Unilateral Actions Taken by One Country", *Journal of Environmental Economics and Management*, 20, pp. 55-0.

Hoel, M. (1992): "Carbon Taxes: An International Tax or Harmonized Domestic Taxes", *European Economic Review. 36*, pp. 400-406.

Mehta, S., *et al.* (1995): *"Controlling Pollution: Incentives and Regulation"*, Sage. Delhi.

Ministry of Environment & Forests, Government of India (1992): *"National Conservation Strategy and Policy Statement on Environment and Developmen"t*, June 1992, New Delhi.

Murty, M.N. (1994): "Management of Common Property Resources: Limits to Voluntary Collective Action", *Environmental and Resource Economics. 4, pp. 581-594.*

Murty, M.N. (1996): "Fiscal Federalism Approach for Controlling Global Environmental Pollution", *Environmental and Resource Economics. 8, pp. 449-459.*

Murty, M. N., A. J. James and Smita, Misra (1999): *"Economics of Water Pollution : The Indian Experience"*, Oxford University Press, New Delhi.

Murty, M.N and Surender Kumar (2004): *"Environmental and Economic Accounting for Industry"*, Oxford University Press, New Delhi.

NEERI, (1992): *"Common Effluent Treatment Plant: State of the Art"*, Nagapur.

Pigoue .A.C. (1920): *"The Economics of Welfare"*, MacMillan, London.

Planning Commission, Government of India, 6th Five Year Plan (1980-85), 7th Five Year Plan (1985-90) and 8th Five Year Plan (1992-97) Documents, New Delhi.

World Bank (1999): *"Greening Industry: New Roles to Communities, Markets and Governments"*, Oxford University Press, New York.

6

Integrated Environmental and Economic Accounting: Framework for a SNA Satellite System

Peter Bartelmus, Carsten Stahmer
and Jan van Tongeren

National accounts have provided the most widely used indicators for the assessment of economic performance, trends of economic growth and of the economic counterpart of social welfare. However, two major drawbacks of national accounting have raised doubts about the usefulness of national accounts data for the measurement of long-term sustainable economic growth and socio-economic development. These drawbacks are the neglect of (a) scarcities of natural resources which threaten the sustained productivity of the economy and (b) the degradation of environmental quality due to pollution and its effects on human health and welfare. In the present paper, the authors attempt to reflect environmental concerns in an accounting framework which maintains as far as possible SNA [System of National Accounts] concepts and principles. To this end, the accounting framework is used to develop a "SNA Satellite System for Integrated Environmental and Economic Accounting" (SEEA). Environmental costs of economic activities, natural asset accounts and expenditures for environmental protection and enhancement, are presented in flow accounts and balance sheets in a consistent manner, *i.e.* maintaining the accounting identities of SNA. Such accounting permits the definition and compilation of modified

Note: The authors thank Hubert Donnevert and Stefan Schweinfest for their assistance on the present version. The views expressed by the authors are their own and not necessarily those of their respective institutions.

indicators of income and expenditure, product, capital and value added, allowing for the depletion of natural resources, the degradation of environmental quality and social response to these effects. A desk study of a selected country is used to clarify the proposed approaches, to demonstrate their application in future country studies and to illustrate the quantitative effects of the use of modified concepts on the results of analysis.

6.1 INTRODUCTION

The discussion of environmentally sound and sustainable socio-economic development has received increased attention by the international community, stimulated in particular by the report of the World Commission on Environment and Development (1987). At its forty-second session, the General Assembly welcomed the Commission's report (resolution 42/187) and adopted an "Environmental Perspective to the Year 2000 and Beyond" which proclaimed "as the overall aspirational goal for the world community the achievement of sustainable development on the basis of prudent management of available global resources and environmental capacities" (resolution 42/186). Environmentally sound and sustainable development will also provide the basic theme for the planned United Nations Conference on Environment and Development in 1992.

The need for clarifying this new development concept and for developing methodologies for its assessment and implementation has been recurrently stressed in international conferences, seminars and workshops. Joint workshops, organized by UNEP and the World Bank, examined the feasibility of physical and monetary accounting in the areas of natural resources and the environment and developed alternative macro-indicators of ecologically adjusted and sustainable income and product (Ahmad, El Serafy and Lutz, 1989). A consensus emerged in the workshops that enough progress had been achieved to link environmental accounting to the standard System of National Accounts, the SNA (United Nations, 1968), and to include certain aspects of environmental accounting in the ongoing revision of SNA.

National accountants and environmentalists reviewed a first draft of the present paper in UNEP/World Bank-sponsored expert meeting (Paris, November 21–22, 1988). The experts at the meeting endorsed the idea of developing a satellite system of environmental accounts and discussed a variety of methodological and procedural questions. These questions should be resolved before preparing an internationally recommended manual of environmental accounting. The experts also requested that the revised SNA should elaborate on the approaches to incorporating environmental concerns in national accounts.

The immediate objective of the present framework is to serve as the basis for preparation of a "SNA Handbook on Integrated Environmental and Economic Accounting" to be issued within the United Nations series of national accounting handbooks. The framework should also facilitate the consideration of environmental accounting in the revised SNA, possibly as part of a more general treatment of the concept of satellite accounts and with appropriate cross-referencing to the Handbook. The draft methodologies have been tested in pilot country studies, and will be distributed widely for comments and contributions.

The framework discussed in this paper is the basic structure for a "Satellite System for Integrated Environmental and Economic Accounting" (SEEA). It is presented in tabular form with an illustrative set of data and is described in some detail in the text. In section 6.2 the main objectives of environmental accounting as well as the general structure of the SEEA are described. Section 6.3 contains a description of the supply side of goods and services, focusing on environmental protection services and the supply of natural growth products. The accounting for the costs of environmental depletion and degradation, resulting from production and consumption, is the main issue of section 6.4. In this section, the authors also explain how these costs affect value added and final demand. One basic indicator, the Environmentally Adjusted Net Domestic Product or "Eco Domestic Product" (EDP) is presented in this context. In section 6.5 the flow accounts of sections 6.3 and 6.4 are complemented by the presentation of stock assets of tangible wealth that include natural assets and changes therein. In section 6.6 the possible extensions of the flow accounts to obtain welfare-oriented macro-indicators are discussed. Finally, some comparative analyses of the conventional and environmentally modified concepts are presented in section 6.7.

6.2 GENERAL FEATURES OF A SATELLITE SYSTEM FOR INTERGRATED ENVIRONMENTAL AND ECONOMIC ACCOUNTING (SEEA)

(a) *Objectives of Integrated Environmental and Economic Accounting*

The focus of traditional systems of national accounts on market and some related non-market transactions (except for imputations for "directly competitive" non-market production of goods and services) has effectively excluded the accounting for changes in the quality of the natural environment and the depletion of natural resources. These effects have been considered to be particularly relevant for the assessment of long-term sustained growth and development and of increases in "social welfare". The overall objective of environmental accounting is thus to measure more accurately the structure, level and trends of socio-economic performance for purposes of environmentally sound and sustainable development planning and policies. The attainment of this objective would facilitate both the systematic compilation and analysis of environmental and related socio-economic data and the formulation of alternative standard macro-economic variables for the analysis of environmental-economic interrelationships.

The current revision of the SNA (United Nations, 1990) presents a unique opportunity to examine how the various concepts, definitions, classification and tabulations of environmental and natural resource accounting can be linked to or incorporated in the SEEA. It may appear premature, however, to radically change a well-established system of economic accounts that serves many different short-, medium-, and long-term socio-economic analyses. Further elaboration of the standards of environmental and natural resource accounting in *SNA satellite system* of environmental accounts has therefore been proposed (Bartelmus, 1987). A similar view was expressed by the experts working on the current revision of the SNA (Lutz and El Serafy, 1989).

Satellite systems of national accounts generally stress the need to expand the analytic capacity of national accounting for selected areas of social concern in a flexible manner, without overburdening or disrupting the "core" system (Lemaire, 1987; Teillet, 1988; Schäfer and Stahmer, 1990). Typically, satellite accounts allow for the:

- provision of additional information on particular social concerns of a functional or cross-sectoral nature,
- linkage of physical data sources and analysis to the monetary accounting system,
- extended coverage of costs and benefits of human activities, and
- further analysis of data by means of relevant indicators and aggregates.

Accordingly, the following specific objectives can be formulated for the planned SEEA:

(i) *Segregation and Elaboration of All Environment-Related Flows and Stocks of Assets of Traditional Accounts:* Satellite accounts, in the narrow sense of detailed accounting for expenditures and revenues in major areas of social concern, were pioneered by France (Institute National, 1986a). There is now an increased interest in segregating all flows and stocks of assets in national accounts related to environmental issues and, in particular, in estimating the total expenditure for the protection or enhancement of the different fields of environment. One objective of this segregation is the identification of the increasing part of the Gross Domestic Product (GDP) which reflects the costs necessary to compensate the negative impacts of economic growth ("defensive expenditures") rather than increases in "true" (welfare-relevant) income (Hueting, Leipert, 1987; Leipert, 1989; and Olson, 1977).

(ii) *Linkage of Physical Resource Accounting with Monetary Environmental Accounting and Balance Sheets:* Physical resource accounts aim at covering comprehensively the total stock or reserves of natural resources and changes therein, even if these resources are not (yet) affected by the economic system.[1] The proposed accounting for these resources is considered the "hinge" by which comprehensive physical resource accounts could be linked to the monetary balance sheet and flow accounts. Another important method for analyzing the environmental-economic interrelationship in physical terms is the development of material/energy balances (Ayres, Kneese, 1969; Ayres, 1978; United Nations, 1976). This approach allows in particular the linkage of input-output tables with data on natural resource inputs, the description of the transformation of natural resources in the production process and the assessment of the generation of residuals of the economic activities (Isard, 1969; Leontief, 1973). Systems of environmental statistics such as those proposed by the United Nations (in preparation) should facilitate achieving compatibility between physical and monetary accounts by specifying those parameters that could be valued in monetary terms to obtain the figures required in environmental accounts. Non-monetary data in physical accounts are considered to be an integral part of the SEEA and will be fully elaborated in the Handbook on Integrated Environmental and Economic Accounting. However, the present framework will concentrate on the monetary stocks and flows of an environmental accounting system.

(iii) *Assessment of Environmental Costs and Benefits:* In contrast to the above-mentioned "narrow" satellite accounts, a broader framework for satellite accounting, covering additional "external" environmental costs and benefits, is proposed here. Taking the current state of knowledge and data availability into account, this framework focuses on expanding and complementing the SNA, with regard to two major issues, namely

- the use (depletion) of natural resources in production and final demand and
- the changes in environmental quality resulting from pollution and other impacts of production, consumption and natural events on one hand and environmental protection and enhancement on the other.

Possibilities of extending the framework for the analysis of environmental welfare effects, *i.e.* the "damage costs" of human health impairment, recreation and other aesthetic or ethical values, are also indicated.

(iv) *Accounting for the Maintenance of Tangible Wealth*: The recent discussion of the new paradigm of sustainable development stressed the need to fully account for the use of both man-made and "natural" capital in order to alert to possible non-sustainable growth and development scenarios. The proposed framework aims at extending the concept of capital to cover not only man-made capital, but also natural capital. Accordingly, SEEA will include additional costs for the depletion and degradation of these natural assets. It will also extend the concept of capital formation to capital accumulation which reflects additionally the deterioration of natural capital as a result of economic uses.

(v) *Elaboration and Measurement of Indicators of Environmentally Adjusted Income and Product:* The consideration of the depletion of natural resources and changes in environmental quality permits the calculation of modified macro-economic aggregates, notably the Environmentally Adjusted Net Domestic Product, short: Eco Domestic Product (EDP).

All these objectives can only be realized step by step. Initial emphasis in practical work should be on the improvement of physical environmental data and on linking them with national accounts as prerequisite for the valuation of environmental effects.

(b) Scope and Structure of the SEEA

The proposed SEEA follows as far as possible the principles and rules established in the SNA (United Nations 1968, 1977, 1990). It is based on SNA's production boundary, follows its analysis of costs and outputs and incorporates the same accounting identities between supply and use of products and between value added and final demand. Information needed for environmental analysis is presented separately. In this manner, original (unadjusted) SNA data can be directly compared with environmentally adjusted statistics and indicators, facilitating the linkage with the central framework of the SNA. Such compliance and linkage with SNA aims at better integration of environmental variables into established economic analysis.

The very nature of a framework allows only the most important concepts and accounting procedures to be highlighted. Definitions, classifications, valuation principles, data sources and processing will be further elaborated on in the Handbook on Integrated Environmental and Economic Accounting. The Handbook will benefit from the experience gained in country studies and existing expertise at the national and international levels.

The present framework seeks to be flexible regarding alternative approaches to integrated environmental-economic accounting and analysis. The interrelationship between the environment and the economy is described as complete as possible. However, in line with the production boundary of SNA, phenomena that take place wholly within the environment, *i.e.* outside the economic system, are excluded. Such phenomena are probably better accounted for by the use of complementary biophysical resource accounts and systems of environment statistics and monitoring. Also, welfare effects from environmental quality degradation that affect "human capital", *i.e.* human health and welfare, are not accounted for in the present framework. However, as shown below (see section 6.6), a "window" to the analysis of environmental damage related to human welfare has been opened, facilitating further extension or alteration of the framework for such analysis.

The main emphasis of the proposed scheme is on the implications of the environment for production, value added, final and intermediate demand and tangible wealth. Therefore, the framework does not present complete accounts for all institutional sectors. Transactions related to income distribution and those concerning intangible assets, including exploitation rights, and also financial assets are excluded. A complete analysis of the interrelationships between the economy and the environment will call for an extended system of all institutional accounts, which shows not only the flows of goods and services, but also of income and finance.

In Table 6.1 the general structure of the system which consists of three basic components is illustrated. In Tables 6.1 and 6.2 the supply and use of goods and services is shown. The asset accounts with opening and closing assets and the items linking them are shown in Table 6.3. Tables 6.2 and 6.3 are connected via the accounts of capital accumulation. The component tables are further elaborated on in Tables 6.2, 6.3 and 6.5 as explained in section 6.3 to 6.5.

The supply Table 6.1 contains an additional row which shows the involuntary "imports" of residuals (wastes etc.) of foreign economic activities which were transported to the domestic economy (−1.6). The use/value added Table 6.2 is extended by row as well as by column. In the table, we show not only the traditional GDP and NDP, but also further corrections due to the use of natural assets (depletion of natural resources, degradation of natural assets by residuals, agricultural and recreational use etc.). This use is valued with the costs which would have been necessary to keep the natural capital intact (ecological valuation; see below section 6.4c for an alternative approach in the case of "exhaustible" resources). These costs are interpreted as the decrease in value of the natural assets comparable to the consumption of man-made fixed assets. The deterioration of the natural assets could be caused by current production activities (59.8), consumption activities (household consumption 17.1) or by (scraps of) produced assets (5.1). The restoration activities of the government diminish the impacts of economic activities on the natural assets (−5.0). The use of natural assets could affect the domestic nature (loss or ecological functions of the produced biological assets −0.9, natural non-produced assets − 73.0) or—as far as the generation of residuals is concerned— could lead to transportation to the rest of the world (exports: −4.7). The value of the deterioration of the domestic as well as foreign natural assets caused by domestic sources (59.8 + 22.2 = 82.0) is used for estimating the environmentally adjusted Net Domestic Product (NDP), called Eco Domestic Product (EDP) (185.1) (see section 6.4c below).

The asset accounts (Table 5.3) show the produced assets (including cultivated biological assets) and the non-produced assets which contain only natural assets (wild biota, land, subsoil assets, water and air). Market valuation is applied except for the depletion and degradation values of natural assets shown in the use/value added table (Table 6.2). These volume changes are valued with the (hypothetical) costs for maintaining them on the same overall quantity and quality level during the reporting period. The question of how such values could be

Table 6.1: System for Integrated Environmental and Economic Accounting (SEEA) (Summary Presentation)

| | | | | | *Tangible assets* | | | *Rest of the World* | |
| | | | | | *Produced* | | *Non-produced* | | |
Use/Value added *(Table 1.2)*	*Total*	*Domestic Production (Industries)*	*Final Consumption* Households	*Government*	*Except Natural*	*Natural (biota)*	*Natural Assets*	*Exports/ Imports*	*Flow of Residuals*
Opening Stocks (Market Valuation)					991.3	83.1	1744.4		
						+ (plus)			
					Capital Accumulation				
					Produced Assets		*Non-Produced*		
					Except Natural	*Natural (biota)*	*Natural Assets*		
Use of products	591.9	224.0	175.0	42.5	68.0	1.4	7.3	73.7	
Gross Domestic Product (GDP)		293.4							
Consumption of fixed capital		26.3			−23.0	−3.3			
Net Domestic Product (NDP)		267.1							
Use of natural assets (ecological valuation)	−1.6	59.8	17.1	−5.0	5.1	−0.9	−73.0		−4.7
Environmental adjustment of final demand		22.2	−17.1	−5.1					
Environmentally Adjusted Net Domestic Product (EDP)	185.1								

(contd.)

Table 6.1 (contd.)

Supply/Origin (Table 6.1)		+ (plus)	
Supply of products	591.9	517.4	74.5
Origin of residuals	− 1.6		− 1.6

		+ (plus)	
Adjustment of natural assets accumulation to market valuation		0.9	81.2
Other volume changes (market valuation)	−25.3		22.8
Revaluation due to market price changes	138.1	12.6	382.8

= (equals)

Closing Stocks (Market Valuation)	1149.1	93.8	2165.5

integrated into the asset balances containing mainly market values is discussed in section 6.5.

6.3 SUPPLY OF GOODS AND SERVICES

The supply table (Table 6.2) includes two elements: gross output, resulting from domestic production, and imports. Gross output is cross-classified by industries and type of product (good or service). Imports are classified by the same type of product as domestic gross output, so that the two elements of supply can be added together to obtain total supply by product. Furthermore, the involuntary "imports" of residuals of foreign economic activities are shown. This item could contain *e.g.* the unaccepted dumping of foreign wastes in national territories.

In Table 6.2 we show a breakdown of domestic production activities by environmental protection activities and other industries. The fully elaborated system will display a further breakdown by industries according to the International Standard Industrial Classification of all Economic Activities (ISIC) (United Nations, 1990a).

A major modification of the SNA is the separate identification of *environmental protection services* from other production activities for all industries. The separation is to facilitate the assessment of the importance of environmental activities in gross output, employment, other production costs, and in capital consumption. Environmental protection services comprise in principle all activities to maintain and enhance the quality of the natural assets. This could be achieved by avoiding environmental impacts of the economic activities (*e.g.* by using integrated or end-of-pipe technologies) or by restoring the natural environment already degraded or depleted. Environmental protection activities can be produced for third parties (external use) as main or secondary production activities of the establishments (36.2) or they can be used internally. The internal provision of environmental protection services is considered to be an "ancillary" activity which is not shown as separate output of the respective establishments in Table 6.2. The cost value of ancillary services is identified separately, however, in Table 5.3– the total of intermediate consumption (17.9), consumption of fixed capital (4.8), compensation of employees (8.7) and net indirect taxes (0.3). These costs are balanced by a negative operating surplus (–31.7). It is not proposed to "externalize" the internal environmental protection activities within the SEEA in order to maintain close linkage with the SNA. For more comprehensive analyses of environmental

Table 6.2: Supply/Origin

| | Domestic Production (industries) | | | | Imports | |
| | | Other Industries | | | | |
	Total	External Environmental Protection Activities	Internal Environmental Protection Activities	Other Activities	Products	Residuals
(1) Supply of products (goods and services)	591.9	36.2		481.2	74.5	
(1.1) Natural growth products	40.7			38.2	2.5	
(1.2) External env. prot. services	36.2	36.2				
(1.3) Other products	515.0			443.0	72.0	
(2) Origin of residuals	-1.6					-1.6
Σ Total supply [(1) +(2)]	590.3	36.2		481.2	74.5	-1.6

expenditures and operations, ancillary activities could be externalized in supplementary tables.

The supply of products is disaggregated in Table 6.2 according to the three categories of natural growth products, external environmental protection services and other products only. A further breakdown of these categories needs to be developed, as far as possible in terms of the Central Product Classification (CPC) (United Nations, in prep.).

Natural growth products of agriculture, forestry and fishing (40.7) refer to those growth-based outputs that are controlled by human activities and can thus be considered as part of planned economic production. Natural growth in these products is treated as primary production which increases the stocks or fixed assets by the amount of growth taking place during the accounting period. On the other hand, those primary natural growth-based products that are largely harvested from the non-controlled natural environment (without human interference in the growth process, such as hunting, gathering of wild fruits, deep-sea fishing or the exploitation of tropical forests) are considered as either "free" inputs or, in case of "scarcities," as environmental depletion costs (see below, section 6.4) of the agriculture, forestry and fishing sectors. For example, in the case of free supply of fish to the fishing industry, the sector's output would not consist of live fish, but rather of fish landed and sold in the market-place.

6.4 USE AND VALUE ADDED

The use/value added table (Table 6.3) shows the use of products and (man-made as well as natural) assets as inputs of the domestic production activities or as components of final demand (final consumption, capital accumulation, exports). These data are supplemented with information on the value added of the different production activities. The use table is an instrument to distribute the total supply of goods and services from the supply table to its various destinations. However, the supply of environmental assets is not displayed in the supply table, but is shown as negative entries in the natural non-produced assets column of capital accumulation. In comparison to the traditional framework of the SNA, the use, of natural assets is shown in additional rows and the capital accumulation of non-produced natural assets in an additional column.

(a) Use of Goods and Services

The first block of rows in Table 6.3 presents the use of products (goods and services) by intermediate consumption of economic activities and final demand, as supplied from Table 6.2 (591.9). This corresponds to the traditional use table in the SNA. The sum of the gross value added (293.4), the conventional Gross Domestic Product (GDP), is shown explicitly in Table 6.3. Subtracting the consumption of fixed capital obtains the Net Domestic Product (NDP) (267.1).

As indicated in section 6.3, the supply of natural growth products (40.7) stems from "controlled" production processes of agriculture, forestry and fishing only. These products are used as inputs into different economic activities (23.0), exported (5.0), consumed by private households (11.3), or may increase fixed capital or stocks (1.4). Stock increase results from the growth in products which are not used in the same period. Stock decrease is shown where naturally grown products of a former period are used for intermediate or final purposes. The increase of fixed capital on the other hand represents a growth in the remaining biomass that is not intended to be used up in intermediate or final consumption, such as the trunks and branches of fruit trees or the breeding stock of livestock.[2]

External environmental protection services (36.2) are used for avoiding potential or restoring actual decreases in environmental quality. It is assumed in the numerical example that the environmental protection services of the government which are not sold on the market (government consumption: 5.0) are restoration activities whereas the other environmental protection activities (31.2) are avoidance activities and are bought by industries (22.4) and households (8.8). Environmental protection activities of the government for avoiding environmental degradation caused by its own production are assumed to be part of the internal environmental protection activities. Government environmental protection services services sold in the market are assumed to intermediate consumption of industries or household consumption.

The other products (515.0) are used for intermediate consumption (178.6), final consumption (192.4), capital accumulation (75.3) and exports (68.7).

(b) Use of Natural Assets

Integrated environmental-economic accounting in the present framework focuses on the inclusion of costs, resulting from the quantitative depletion of natural resources and from the qualitative degradation of environmental quality by economic activities.

Depletion activities (at a total of 18.2) are shown in Table 6.3 to

Table 6.3: Use Value Added

		Domestic Production (industries)				Final Demand							
			Other Industries			Final Consumption		Produced Assets		Net Capital Accumulation			
	Exports Total	External Environmental Protection Activities	Internal Environmental Protection Activities	Other Activities	Subtotal Domestic Production	Households	Government	Except Natural	Natural (biota)	Non-Produced Natural Assets	Products	Residuals	Subtotal Final Demand
(1) Use of products	591.9	15.9	17.9	190.2	224.0	175.0	42.5	68.0	1.4	7.3	73.7		367.9
(1.1) Natural growth products	40.7			23.0	23.0	11.3			1.4		5.0		17.7
(1.2) External envir. protection services	36.2			22.4	22.4	8.8	5.0						13.8
(1.3) Other products	515.0	15.9	17.9	144.8	178.6	154.9	37.5	68.0		7.3	68.7		336.4
Gross value added of industries [(9)−(1)]	267.1	20.3	−17.9	291.0	293.4			−23.0	−3.3				−26.3
(2) Use of produced fixed assets (consumption of fixed capital)	0	1.3	4.8	20.2	26.3			−23.0	−3.3				−26.3
Net value added of industries [(9)−(1)−(2)]	267.1	19.0	−22.7	270.8	267.1								
(3) Use of natural assets (ecolog. valuation)	−1.6	6.3	4.6	48.9	59.8	17.1	−5.0	5.1	−0.9	−73.0		−4.7	−61.4
(3.1) Quantitative depletion	0	0.3	0.4	16.8	17.5	0.7			−0.9	−17.3			−17.5
(3.2) Degradation of land (except by residuals)	0	0.2		8.8	9.0	0.8				−9.8			−9.0
(3.3) Degradation by residuals	−1.6	5.8	4.2	23.3	33.3	15.6	−5.0	5.1		−45.9		−4.7	−34.9
Σ Total use [(1) + (2) + (3)]	590.3	23.5	27.3	259.3	310.1	192.1	37.5	50.1	−2.8	−65.7	73.7	−4.7	280.2
(4) Environmental adjustment of final demand	0	1.8	2.1	18.3	22.2	−17.1	−5.1						−22.2

(contd.)

Table 6.3 (contd.)

	c1	c2	c3	c4	c5	c6	c7	c8
Env. adj. net value added (EDP) of industries [(9) − (1) − (2) − (3) − (4) or (5) + (6) + (7) + (8)]	10.9	−29.4	203.6	185.1	37.5	45.0	73.7	258.0
(5) Compensation of employees	13.0	8.7	72.0	93.7				
(6) Indirect taxes minus subsidies	2.0	0.3	34.1	36.4				
(7) Net operating surplus	4.0	−31.7	164.7	137.0				
(8) Eco-margin [−(3) − (4)]	−8.1	−6.7	−67.2	−82.0		−2.8		
(9) Total gross inputs/total final demand [(1) + (2) + (3) + (4) + (5) + (6) + (7) + (8)]	36.2	36.2	481.2	517.4	175.0	−65.7	−4.7	

consist of depletion of natural assets by industries (17.5) and by households (0.7). As detailed in Table 6.5, they comprise the exploitation of natural resources such as sub-soil assets (mineral deposits) by mining and quarrying (–8.9), aquifers (–4.7) and biological assets (*e.g.* timber from tropical forests or fish stocks of inland and marine waters) by agriculture, forestry and fishing (–0.9, –3.7). The assumption is that scarcities in the availability of renewable (forest, fish, wildlife etc.) and cyclical (water) resources have been observed. Depletion costs are only estimated in these cases as far as the economic use of natural assets leads to imbalances in nature, *i.e.* if the depletion of biota exceeds the natural growth or the use of water exceeds replenishment of aquifers. The recording of corresponding negative amounts of tangible wealth reduction is discussed below in section 6.4d.

The other category of economic use of natural assets represents the environmental quality degradation of the environmental media of air, water and land by production and consumption activities. The degradation of land could be caused by improper agricultural practice (soil erosion, water logging, salinization), by excessive use for recreational purposes or by polluting the soil with wastes or waste-water. The main reason for degrading the quality of air and water is their use as a sink for residuals (wastes, pollutants) of economic activities. It has to be stressed that only the immediate influence on the environmental media is taken into account. The indirect effects by transboundary transport in the air or by transition from one environmental medium to another are not recorded in the SEEA. These complex dynamics within the natural environment could be shown in supplementary data systems which should be linked with the SEEA. Furthermore, it should be noted that impacts of natural or man-made disasters are assumed not (or in some cases only indirectly) to be caused by economic use of environmental assets and are therefore excluded from the use/value added Table 6.3 but are included as a category (4) of asset volume changes in Table 6.5.

The net value of *degradation* is assumed to be equal to potential abatement (restoration) costs, required either to achieve the level of environmental quality at the beginning of the accounting period or at least a level specified by "official"environmental standards (Hueting, 1980). It is assumed that such standards reflects a technological solution to abating environmental quality degradation that can "reasonably" be expected to be applied by the different polluters. Obviously, such valuation does not measure actual environmental "damage" from pollution. A possible treatment of such welfare effects is discussed in section 6.5.

The environmental degradation is caused by production activities (9.0 plus 33.3), by consumption activities of households (0.8 plus 15.6), by man-made assets (5.1) and by imported residuals (–1.6). Man-made assets have an effect on the natural environment by their residuals (e.g. scrapped machinery). A part of the environmental degradation is restored by government activities–5.0). The remaining degrading impacts affect the domestic natural assets(–9.8, –45.9) and –as far as residuals are "exported"–the natural environment of the rest of the world (–4.7).

In Table 6.4, the value of the economic use (depletion as well as degradation) of domestic and non-domestic (foreign) natural assets and the corresponding impacts on the asset values are shown in a simplified balance sheet.

Table 6.4: Economic Use and Impacts on Natural Assets

Use of Natural Assets (environmental costs)		Impacts on Natural Assets (decrease of asset values)	
Domestic use		**Domestic environment**	
Depletion		*Depletion*	
industries	17.5	prod. natural assets	0.9
households	0.7	non-prod. natural assets	17.3
	18.2		18.2
Degradation		*Degradation*	
industries	42.3	non-prod. natural assets	55.7
households	16.4		
government	– 5.0		
prod. assets	5.1		
		Environment of the	
Imports		**rest of the world**	
Degradation	1.6	*Degradation*	4.7
	60.4		60.4
	78.6		78.6

(c) Environmental Adjustment of the Value Added

Deducting the imputed costs of natural asset use (environmental costs) from net value added leads to a new value-added concept, termed here "environmentally adjusted net value added." The environmental costs represent the hypothetical costs for maintaining the natural assets at the same level during the reporting period. This concept reflects a "strong" or "narrow" sustainability concept which implies that future generations should receive a natural environment with a quantitative and qualitative level being at least comparable with the present situation (Bartelmus, in preparation: Blades, 1989; Daly, in preparation; Pearce, Markandya, Barbier, 1989 and 1990; and Pezzey, 1989). The international discussion of the last years has proved that it is not sufficient to sustain a constant level of total (man-made as well as natural) capital, denying substitution possibilities between these capital categories ("broad" or "weak" sustainability concept). The uncertainty of long-term impacts of economic activities on the natural environment and the increasing knowledge about irreversible damages of natural balances (climate change, ozone layer depletion etc.) has led to a more cautious risk-conscious attitude towards overburdening the natural environment. From this point of view, it seems necessary to maintain the natural assets treating them as complementary to man-made capital. The strong-sustainability concept thus applies not only to the case of environmental quality degradation, but also to the maintenance of "stocks" of natural resources. In the case of subsoil assets, this approach seem to be questionable because the strong sustainability concept would lead to non-use of the resources, possibly causing severe world-wide economic problems. Instead, the objective could be to maintain a long-term optimal depletion rate, considering that new finds could only retard the shrinkage of the stocks. It has been proposed that the sustainability concept should be weaker in this case, and it would be sufficient to balance a decrease of the subsoil assets with an increase of other types of assets (with preference for permanent or renewable natural assets) to sustain the same income level in the future (El Serafy, 1989; Daly, in preparation).

The maintenance cost approach used for valuing the economic use of natural assets corresponds to the methods of national accounting for estimating the use of man-made fixed assets. The user costs of these assets are estimated with the costs necessary to keep the man-made fixed capital intact, *i.e.* to maintain the level of the assets at the same level during the reporting period. These costs which are called "consumption of fixed capital" or "depreciation" are also used to compile the net capital formation of the man-made assets in the accounting period.

As far as the natural assets have the character of *fixed assets*, treating the maintenance costs of natural assets in the same way as the depreciation of man-made assets seems plausible. However, distinguishing between assets that bear characteristics of fixed assets and those that are more in the nature of an *inventory or stock* (in this case, decrease of assets in the national accounts is booked as intermediate consumption and not as depreciation) is problematic because natural assets may exhibit simultaneously economic and environmental functions (Hueting, 1980). For instance, a timber tract represents a stock resource, but has also an important role in cleaning the air and regulating water balances.

Furthermore, it serves as habitat of animals and as recreational area. From an ecological point of view, the environmental media, *i.e.* land, water and air as well as the ecosystems can be considered as fixed assets. The maintenance costs of these assets should therefore be treated as depreciation. Further discussion seems to be necessary in the case of subsoil assets. They mainly have the character of inventory stocks of nature. Their depletion could therefore be treated as intermediate consumption.[3] For sake of simplicity of the present framework, the value of the depletion of these assets is not shown separately from the other environmental costs, but is also treated as decrease of a fixed asset.

Whatever the treatment of the environmental costs, as depreciation of natural assets or as intermediate consumption, their deduction from gross output affects the calculation of net value added. The gross value added of the industries remains unchanged in the SEEA. The environmental adjustments of the net value added (−82.0) comprise the imputed environmental costs connected with domestic production (−59.8), household consumption activities (−17.1) and the use of man-made assets (−5.1). These adjustments are called eco-margin which is introduced explicitly in order to permit the identification of all components of value added (including operating surplus) according to the conventional SNA concepts.

Impacts of household activities and of man-made **assets** on environmental quality are taken into account for correcting the net value added despite the fact that the respective environmental costs are not directly associated with production activities. Regarding households, their polluting activities could be viewed as non-market production of goods and services which produces "jointly" residuals like wastes and pollutants. In this case, the net value added of the households' production would be diminished by the imputed environmental costs of the households. This is achieved by shifting these imputed values (17.1) from final consumption to the totals of domestic production. A similar correction is made with regard to the environmental impacts of man-made assets, comprising additional imputed costs of the asset owners (5.1). These costs refer *e.g.* to pollution caused by controlled landfill and to the residues of unrecycled man-made assets. It is theoretically possible to transfer these costs to the different industries. In this case, their net value added would directly be affected. This procedure has not been applied in the SEEA in order to show separately the environmental costs caused by current production and man-made capital use. The shift of the environmental costs of households and man-made assets to the columns of domestic production is shown in Table 6.3 in the row "environmental adjustments of final demand." Net value added is thus corrected only for the totals of environmental protection activities (1.8 and 2.1) and of other activities (18.3)

In Table 5.3 we also record the components of value added, consisting of the compensation of employees, indirect taxes net of subsidies, net operating surplus and environmental costs equal to item (8) "eco-margin." Use of SEEA thus permits the analysis of these components of value added for the environmental protection activities of the different economic sectors. Indirect taxes, and subsidies, charged or granted as part of environmental protection policies, will be identified separately in the SEEA, reflecting the application of polluter-and user-pays-principles at the micro-economic level. Macro-policy makers on the other hand, might be concerned with the assessment of employment devoted to "defensive" environmental protection activities (total "environmental" renumeration of employees of 21.7 as compared to a total of other wages and salaries of 72.0).

The net operating surplus of the different production activities has not been environmentally adjusted in Table 6.3. The additional environmental costs are balanced by introducing the eco-margin. The idea is to facilitate the unequivocal linkage of the production accounts of the SEEA with the income accounts of the conventional SNA. Another possible presentation of the operating surplus could show an environmentally adjusted net operating surplus, but extend the table at the same time for identifying explicitly the non-adjusted gross and net operating surplus:

Environmentally adjusted net operating surplus	55.0
= Gross operating surplus	163.3
− Consumption of fixed capital	26.3
− Eco-margin	82.0

The total of the environmentally adjusted net value added is called Environmentally Adjusted Net Domestic Product or, short, Eco Domestic Product (EDP). EDP could be derived from GDP as follows:

Gross Domestic product (GDP)	293.4
− Consumption of fixed capital	26.3
= Net Domestic Product (NDP)	267.1

$$- \text{(Imputed) environmental costs} \quad 82.0$$
$$= \text{Eco Domestic Product (EDP)} \quad 185.1$$

(d) Final Demand

Final demand consists of final consumption, net capital accumulation and exports. Import and export flows are only slightly modified for environmental accounting. However, significant alterations are proposed for both final consumption and net capital accumulation to allow corrections of net value added while heeding the principle of accounting identities (see below, section 6.4e).

Imports and exports include flows of wastes which are not marketed, but transported to/from a foreign country or to the open sea. They represent either a degradation of the foreign natural media by exporting domestic residuals or of domestic media by importing residuals. They are estimated as negative values—costs for avoiding or restoring environmental quality degradation (exports:–4.7, imports:–1.6). The imports of residuals reduce the total value of imports and of the domestic natural assets (negative item in the column of net accumulation of natural assets). The exports of residuals lead to an increase of the imputed environmental costs of the exporting industry which implies a reduced environmentally adjusted net value added of the exporting economic units and a reduction of the total value of the exports. Transboundary flows of residuals of the economic activities which are not transported by man but by environmental media (*e.g.* water, air), are recorded as degradation of the environmental media which directly receive the residuals. Their final destination is not taken into account.

The conventional final consumption of households (175.0) remains unchanged in SEEA. The additional (imputed) environmental costs (17.1) which would have been necessary to avoid or which need to be incurred in order to restore a degradation of environmental quality by household activities (recreactional use of land 0.8, pollution 15.6), or which represent the costs of depleting natural resources (firewood consumption 0.7) are shifted to domestic production.

The fully elaborated SEEA will comprise a further breakdown of the final consumption of households by environmentally oriented functions for identifying *e.g.* the environmental protection expenditures of the households and the expenditures required to compensate for the damages caused by environmental deterioration (health expenditures etc.).

Final *consumption of government* (42.5) is corrected by the environmental protection expenditures (–5.0) which are non-marketed and which are undertaken to avoid or restore a decrease of environmental quality caused by other economic units. These expenditures have the characteristics of an investment in environmental quality. Its value is shifted from the government final consumption to the capital accumulation of natural assets and diminishes the degradation of the natural assets which would have occurred if no restoration activities had taken place. The environmental protection activities of the government for own purposes (internal activities) and the additional (imputed) environmental costs of the government production activities are already recorded in the columns of domestic production. It is therefore not necessary to extend the concept of final consumption of government in the SEEA by taking into account imputed environmental costs.

The section on the *net accumulation of tangible wealth* in Table 6.3 differs considerably from the traditional incorporation of capital formation in a use table. The presentation of this part of final demand in Table 6.3 is limited to an asset classification by only three types of asset: produced biological (natural) assets, other produced assets, non-produced natural assets. A further breakdown of the capital accumulation is given in Table 6.5 which shows complete asset balances for the different types of assets. The following comments refer especially to the disaggregated version in Table 6.5. The capital accumulation concept of Table 6.3 corresponds to the (traditional) capital formation (item 2 of Table 6.5) and to the ecological valuation of volume changes of natural assets due to economic use (item 3.1 of Table 6.5). The valuation problem (market versus ecological valuation) and the other items of Table 6.5 are discussed in section 6.5.

In the SEEA, the asset boundary has been extended for including all natural assets which are actually or potentially used by economic activities or which could be affected by the residuals of economic production and consumption processes. The extended asset concept comprises the following types of assets:

Produced assets
>   Man-made assets (non-biological such as machinery and equipment, stocks of non-biological products)
>   Natural assets produced by agriculture, forestry and fishing (fixed assets and inventory stocks)

Non-produced natural wild assets

>   Wild biological assets
>   Land (cultivated and uncultivated) ·
>   Subsoil assets (developed and undeveloped proven reserves)

Water (stored and unstored)

Air

This classification distinguishes in particular between assets which are (economically) produced or not and which are man-made or natural. These two criteria are not identical because the (economically) produced biota are both produced and natural. In this case, the produced biota should only be subsumed under natural assets as far as they are living. A further breakdown is possible according to the degree of human influence on the natural environment (*e.g.* cultivated uncultivated land, developed-undeveloped subsoil assets).

In Table 6.3, the net capital accumulation of *produced assets* is recorded mainly according to the conventional concepts of the SNA (gross capital formation: 68.0 and 1.4, consumption of fixed capital:– 23.0 and –3.3). Only two minor deviations should be mentioned: The residuals of the produced assets which are loaded on the natural environment (*e.g.* scraps, pollution of controlled landfill) are valued with their avoidance costs. (5.1) and shown in addition to the net capital formation. In a second step, these imputed costs are shifted to the industries of the responsible activities or, alternatively, to the industries as a whole [via the environmental adjustment row (4)]. In the case of produced biological assets, it might be necessary to estimate additional depletion costs (–0.9) if the economic activities of agriculture, forestry and fishing disturb the natural balances, *e.g.* if the amount of cut wood exceeds the natural growth and destroys the ecosystems of cultivated forests. In this case, the sustainability principle should be applied, and avoidance or restoration costs could be calculated.

The imputed depletion costs of *wild biota* (in Table 6.5: –3.7) are estimated only if depletion by economic activities (*e.g.* hunting, ocean fishing) and natural growth are not balanced. Depletion costs are thus estimated if depletion exceeds natural growth. The discussion of valuation of net depletion in this case has not been conclusive. One possible approach could be to value net depletion as the gross value added generated by the depleting activity. This would show value added foregone if the net depletion had been avoided. Another approach could be to assess the costs for compensating projects to restore the natural balances.

Net capital accumulation of *land* refers to the impacts of economic land use. The costs of developing land are treated in the conventional SNA as capital formation which normally leads—from an economic point of view—to an improvement of land quality and to increasing market values (in Table 6.5: 4.6).

Table 6.5: Asset Balances of Net Tangible Assets

Monetary units

| | | Produced Assets | | Non-produced Natural Assets | | | | | |
		Produced Assets (except biological)	Produced Biological Assets	Non-produced Biological Assets	Land (landscape, ecosystems) Cultivated etc.	Uncultivated	Sub-soil Assets	Water	Air
	Total								
(1) Opening stocks (market values)	2818.8	991.3	83.1	65.4	1366.7	50.4	261.9		
(2) Net capital formation (use of products, market values)	50.4	45.0	− 1.9		4.6		2.7		
(2.1) Gross capital formation	76.7	68.0	1.4		4.6		2.7		
(2.2) Consumption of fixed capital	− 26.3	− 23.0	− 3.3						
(3) Volume change of natural assets due to economic use (market values)	36.0			− 2.1	23.3	− 5.0	19.8	0.0	0.0
(3.1) Ecological valuation									
(3.1.1) Quantitative depletion	− 18.2		− 0.9	− 3.7			− 8.9	− 4.7	
(3.1.2) Degradation of land (except by residuals)	− 9.8				− 2.1				
(3.1.3) Degradation by residuals	− 45.9			− 7.7	− 9.5	− 3.1		− 12.9	− 20.4
(3.2) Adjustment due to market valuation									
(3.2.1) Quantitative depletion	8.1		0.9	1.6			0.9	4.7	
(3.2.2) Land use (except by residuals)	35.2				33.1	2.1			
(3.2.3) Degradation by residuals	38.8				4.0	1.5		12.9	20.4

(contd.)

Table 6.5 (contd.)

(3.3) Other volume changes (change of land use, new finds, new estimates etc.)	27.8				3.4	− 3.4	27.8
(4) Volume change by natural or multiple causes (market values)	− 30.3	− 25.3		1.3	− 4.3	− 2.0	
(5) Revaluation due to market price changes	533.5	138.1	12.6	11.1	331.0	11.8	28.9
(6) Closing stocks (market values) $[(1)+(2)+(3)+(4)+(5)]$	3408.4	1149.1	93.8	75.7	1721.3	55.2	313.3

From an ecological point of view, increasing economic use of land could cause a qualitative degradation of the land and the terrestrial ecosystems. The main reasons are restructuring (further economic development of cultivated land, cultivation of uncultivated land), intensive agricultural use (soil erosion etc.), recreational use (disturbing ecosystems) and use as sink for residuals (such as pesticides and the pollution of controlled and uncontrolled landfill). In Table 6.5 degradation by residuals (−9.5 and −3.1) and by other economic activities (−7.7 and −2.1) are distinguished.

The degradation of land is valued as the cost to avoid (or at least mitigate) the negative impacts of economic activities or to restore the degraded areas, with a view to maintaining the terrestrial ecosystems in their present state. This valuation concept might differ widely from the market valuation. Changes in economic land use will often increase the market value of land, but at the same time could imply a decrease of the ecological quality of land.

Subsoil assets comprise the proven reserves of fossil and mineral assets. Proven reserves normally have to meet three criteria: high probability of existence (95 per cent), exploitability with existing techniques and positive net return, *i.e.* the market price exceeds exploitation costs (Martinez *et al.*, 1987). Subsoil assets can be undeveloped or developed (established mines and other exploration facilities). The costs for developing subsoil assets (*e.g.* by exploration activities) have to be treated according to the conventional SNA as capital formation (Table 6.5:2.7). The valuation of the depletion of subsoil assets has to reflect the future scarcity of the assets. Exploitation is mainly an economic and not an ecological problem because the immediate impacts on natural balances are usually low, with the notable exception of surface mining. The indirect impacts of subsoil depletion (*e.g.* losses of crude oil during transportation, pollution connected with energy consumption) are registered independently from the valuation of assets as environmental degradation from polluting economic activities.

Various methods have been proposed to value subsoil asset depletion (see Ward, 1982). Several authors suggest the use of the net operating surplus of the exploiting industry or a part of it. The proposal of El Serafy (1989) seems to be an approach tailored to the concept of sustainability. The idea is to estimate the depletion costs as the amount of money which should be invested to achieve a long-term constant flow of income, even after complete exploitation of the resources. This rule implies a substitution of the use of subsoil assets by other types of income generating activities and corresponds to a broad sustainability concept. The decrease of subsoil assets could be balanced, *e.g.* by increasing renewable (biological) assets or by the development of solar and wind energy sources instead of coal of crude oil. The value of subsoil depletion amounts to 8.9, in Table 5.5.

The economic use of *water* could lead to increasing scarcity (depletion) or to decreasing quality (degradation by residuals). Increasing scarcity of water will be observed if the economic abstraction exceeds the average natural inflow of water during the accounting period. In this case, net depletion could be valued as the value added, or part of it, generated by additional water use of the abstracting industries (−4.7). This value could represent the avoidance costs as in the case of wild biota. Further discussion is necessary to develop a generally accepted valuation method for water depletion. The degradation of water by residuals is valued as its avoidance (or restoration) costs (−12.9).

As described above (section 6.4b), the value of the degradation of *air* is its avoidance costs (−20.4).

In section 6.5 below, we describe a comprehensive system of balance sheets of tangible assets, including changes in volume not accounted for by capital accumulation in the use/value added table (*e.g.* effects of natural and other disasters).

(e) Accounting Identities

The national accounting identities between the totals of environmentally adjusted value added (plus imports) and final demand are maintained in the use/value added Table 6.3 by treating capital accumulation of natural assets as part of final demand. In Table 6.6 we show the transition from the conventional aggregates, according to the SNA, to the environmentally adjusted aggregates of the SEEA by using the numerical example (cf.Tables 6.2 and 6.3).

As already explained, the gross value added (293.4) is corrected by the consumption of fixed produced capital (26.3) and by the user costs of natural assets (current production 59.8, households 17.1, residuals of man-made assets 5.1). This obtains the environmentally adjusted net value added (185.1). The concept of imports is extended to additionally include imports of residuals (74.5 and −1.6). The total of the environmentally adjusted primary inputs (value added plus imports) is 258.0

Domestic final demand (294.2) is corrected by the consumption of fixed produced assets (26.3) to achieve a net concept. The environmental restoration costs of the government (5.0) are treated as an increase of

Table 6.6: Accounting Identities

Primary Inputs *(value added, imports)* Gross value added		Final Demand *(domestic, exports)* Domestic final demand	
(Gross domestic product)	293.4	(SNA concept)	294.2
– Use of produced assets (consumption of fixed capital)	26.3	– Consumption of fixed capital	26.3
– Use of natural assets for current production	59.8	– Government restoration costs	5.0
Environmental adjustment of final demand	22.2	+ Net capital accumulation of natural assets (-78.9+5.0)	– 73.9
Environmentally adjusted net value added (Environmentally adjusted Net Domestic Product)	185.1	Environmentally adjusted domestic final demand	189.0
+ Import of products	74.5	+ Export of products	73.7
+ Import of residuals	– 1.6	+ Export of residuals	– 4.7
Environmentally adjusted primary inputs	258.0	Environmentally adjusted final demand	258.0

the value of natural assets and therefore reflected in the capital accumulation of natural assets. The value of depletion and degradation of natural assets by economic activities would have been –78.9 without government restoration activities. Taking these activities into account, the net capital accumulation amounts to –73.9 (–73.0 plus –0.9). The environmentally adjusted figures of total final demand (258.0) comprise the adjusted domestic final demand (189.0) and the export of products (73.7) and residuals (–4.7). This total is equal to the total of primary inputs.

6.5   ASSET BALANCES OF TANGIBLE WEALTH

As illustrated in Table 6.1, the section of the use/value added table on tangible wealth accumulation can also be viewed as an integral part of the asset balances. This is indicated in Table 6.1 by plus (+) and equal (=) signs, inserted between the four elements of the asset balances. This illustration shows an accounting identity between the closing stocks and the sum of opening stocks, net capital accumulation, adjustment of natural assets accumulation to market valuation, other volume changes and revaluation due to market price changes. This identity holds for man-made assets (produced, not natural):

$$1,149 = 991.3 + (68.0 - 23.0 + 5.1 - 5.1) + (-25.3 + 138.1),$$

for the (economically) produced natural assets:

$$93.8 = 83.1 + (1.4 - 3.3 - 0.9) + (0.9 + 12.9),$$

and for those non-produced natural assets which are used or affected by economic activities:

$$2,165.5 = 1,744.4 + (7.3 - 73.0) + (81.2 + 22.8 + 382.8).$$

The asset balance sheets are further elaborated on in Table 6.5. The asset classification has already been described in Section 6.4d. The volume and price changes of the assets during the reporting period are further disaggregated in Table 6.5, consisting of:

(2) Net capital formation (use of products);

(3) Volume change of natural assets due to economic use;

(3.1) Ecological valuation,

(3.2) Adjustment due to market valuation,

(3.3) Other volume changes (market valuation),

(4) Volume change by natural or multiple causes;

(5) Revaluation due to market price changes.

The volume changes (2) and (3.1) reflect the net capital accumulation described in the use/value added table (Table 6.3 and Table 6.2), the adjustment due to market valuation (3.2) corresponds to the "adjustment of natural assets accumulation to market valuation" in Table 6.3. The other volume changes due to economic use (3.3) and the volume change by natural or multiple causes (4) are summarized under "other volume changes" in Table 6.3. The revaluation due to market price changes is presented in both tables under the same name.

The design of the asset balance sheets aims at introducing environmental aspects in the national stock accounts without disrupting the concepts of the conventional SNA balance sheets. As recommended in the International Guidelines on Balance-Sheet and Reconciliation Accounts (United Nations, 1977) and in chapter XI of the preliminary draft of the revised SNA (United Nations, 1990), the *opening* and *closing stocks* are valued at market prices or have values derived from market prices. Direct market valuation could be applied if the assets are marketed (some produced fixed assets like cars, inventory stocks of products, land). Indirect market valuation uses the net value concept (replacement costs minus cumulated depreciation) or tries—in the case of depletable natural assets like wild biota, subsoil assets or water—to estimate the assets by the discounted value of future net returns. (future market prices minus all exploitation costs including a normal rent of capital).[4] It should be stressed that the SEEA does not aim at a complete market valuation of the non-produced natural assets. The market valuation should be limited to natural assets which are regularly depleted for market purposes (*e.g.* ocean fish, tropical wood and subsoil assets) or to assets which are directly marketed (uncultivated land in exceptional cases). The opening and closing stocks of the other non-produced natural assets have a market value of zero. In these cases, their volume changes are valued only if they are affected by economic activities.

Market valuation has also been applied in general for the *volume* and *price changes* during the accounting period. Net capital formation of produced assets [item (2) in Table 6.5] reflects the volume changes described in the conventional SNA framework. Some of the other volume changes of assets caused by economic, natural, non-economic and multiple (combination of these causes) activities and events [items (3) and (4) in Table 6.5] which had been part of the reconciliation accounts in the 1968 version of the SNA, will be integrated into the accumulation accounts which explain the changes in the balance sheets of the revised SNA at market values. Opening and closing stocks of these assets are also measured at market values. The transition to the level of the market values at the end of the accounting period (closing assets) is shown as item (5) in Table 6.5 (revaluation due to market price changes).

The connection between the SEEA and the conventional assets balance sheets is introduced by a breakdown of the volume change of natural assets due to economic use (item (3) of Table 6.5). As far as the economic use affects the natural balances and leads to a decrease of the value of the natural assets from and ecological point of view, the avoidance or restoration costs are estimated for maintaining the same qualitative and quantitative level of natural capital during the accounting period. These values are introduced in the extended use/value added table of the SEEA (see Table 6.3). This ecologically oriented valuation does not necessarily correspond to the market values due to the respective economic use. Therefore, an adjustment item is introduced which allows the transition to the market valuation of the asset balance sheet [item (3.2) of Table 6.5]. Volume changes due to economic activities which do not directly deplete or degrade the natural assets (changes in land use, discoveries, etc.) are separately recorded and have market values [item (3.3) of Table 6.5]. In the SEEA, the analysis of the volume changes of the natural assets is focused on the economic uses. Therefore, no ecologically orientated valuation of the volume changes of natural assets due to natural or multiple causes (like wars and disasters) is applied, but market values of the volume changes are given.

It is not possible to describe the volume changes of the different types of assets in this overview article in detail. More detail on the extended asset balance sheets will be given in the SNA Handbook on Environmental Accounting. The following limited observations are thus only to facilitate a better understanding of the general scope and coverage of environmental assets in the SEEA.

The consumption of fixed capital (fixed produced assets) comprises only insurable risks of premature losses. Further losses by war or natural disasters are recorded under item (4) (−25.3).

The asset balances of the *biological assets* are relatively complicated because of the different concepts of describing the volume changes of produced and non-produced biota at market values. The natural growth of produced biota is treated as economic production (and gross capital formation) whereas the natural growth of non-produced biota is, as far as market values are associated to them, part of "(4) Volume change by natural or multiple causes" of Table 6.5. The depletion (due to economic use) of the produced biota is shown as decrease of stocks or consumption of fixed capital [item (2) of Table 6.5], whereas the depletion of non-produced biota is indicated under item (3). This different treatment implies that the net growth (natural growth minus depletion) of produced biota at market values is shown as net capital formation under item (2) (−1.9 = 1.4–3.3), while the net growth of non-produced biota equals the difference of the values under item (4) and item (3) (−0.8 = 1.3 − 2.1).

The ecological value of depleting produced biota (−0.9) reflects the ecological consequences of depleting cultivated biota beyond the economic use of these assets. If, for example, the wood of timber tracts is cut, the natural balance of forest could be disturbed as far as depletion exceeds natural growth. The necessary "ecological" costs could be estimated by the costs of compensating projects or by the additional value added generated by the net depletion (gross depletion minus natural growth). Further considerations are necessary to avoid double-counting if the depletion of produced biota exceeds their natural growth. In this case, the ecological valuation has to take into account that the (negative) net growth has already been valued at market values as (negative) capital formation.

The depletion of wild biota (−3.7) could be valued in ecological terms in a similar way. The natural balances could only be maintained if depletion and natural growth are balanced. That is, if the net depletion is positive avoidance or restoration costing refers to a reduction of the production (hunting, harvesting etc.) and a decrease of the corresponding value added, and the loss of ecological functions of the resource.

The valuation of the quality changes of *land* caused by economic activities might produce completely opposite results, depending on the economic or environmental point of view. Restructuring and development of land are normally connected with increasing market values, whereas their ecological effects could decrease the land values under environmental aspects. The development costs (4.6) are shown as capital formation. They reflect, together with the market value of the volume changes due to economic use (23.3,–5.0), the market value of all quantitative and qualitative volume changes of land caused by the different economic activities. The quantitative aspects of land use (changes in land use) are described under item (3.3) of Table 6.5 (3.5, –3.4). The qualitative component is shown under ecological aspects first (–7.7. –9.5, –2.1, –3.1) and adjusted to market valuation in a second step (33.1, 4.0, 2.1, 1.5). The qualitative changes do not only comprise the results of restructuring and development, but also excessive economic use, *e.g.* for agricultural purposes (often connected with soil erosion) and for recreation. Furthermore, the degradation by residuals is taken into account.

The ecological valuation of land degradation raises difficult estimation problems. In principle, the adequate avoidance or restoration costs to maintain the same level of land quality has to be estimated. Avoidance costs could comprise the decrease of value added in case of reducing excessive land use. Restoration costs could be the costs of compensation projects.

The opening and closing stocks of *subsoil assets* (proven reserves: developed and undeveloped) are valued with the discounted value of future net returns (*i.e.* revenues minus exploitation costs: 261.9, 313.3). New discoveries and changes in the economic conditions of exploitation which lead to new estimates of the proven reserves, are shown under item (3.3) of Table 6.5 (27.8). The exploitation costs do not contain the exploration costs (2.7) because they have already been included under item (2) to Table 6.5. The extraction (depletion) of these assets is estimated at "ecological" values (–8.9) and in a second step adjusted to market values (–8.9 + 0.9 = –8.0). These market values reflect the net prices of the depleted assets (current market price minus exploitation costs). The ecological valuation could comprise the costs for maintaining the level of natural capital (compensating projects to develop renewable or permanent assets) or of the total capital (man-made and natural).

The stock of *water* has normally no market values. Exceptions are stored water for drinking or irrigation purposes. The depletion of water is valued from an ecological point of view only if the average water stock is affected. This net depletion is valued with its avoidance costs [costs of reducing water use, *e.g.* by reducing agricultural production (−4.7)]. The avoidance cost approach can also be applied for valuing water degradation by residuals (−12.9).

The *air,* as natural asset, has no market value. Therefore, the value of opening and closing stock is zero. For balancing the value of degradation by residuals (−20.4), a corresponding positive item has been introduced as an adjustment to market valuation.

6.6 WELFARE-ORIENTED MEASURES OF THE ECONOMIC USE OF THE ENVIRONMENT

The concept of sustainability used in this paper is cost-oriented rather than welfare-oriented.[5] It reflects cost estimates which would be necessary to avoid, restore or replace decreases of environmental quantities and qualities during the reference period. Such an approach would normally suggest a greater effort at protecting the environment, as compared to estimating an economically optimal level of pollution. Optimality would require a balance of marginal costs of protection activities and of the (discounted) flows of marginal future environmental damages avoided. Because of underestimation uncertainty and undervaluation (high discounting) of future damage, the optimality criterion will almost certainly present an amount of environmental deterioration which might be optimal from a micro-economic point of view, but not from a social point of view. In view of the uncertainties related to individual (marginal) evaluation and of prevailing societal and international concerns over long-term threats to critical life-support systems, the cautious concept of sustainability implicit in the cost values of the present framework, *i.e.* the maintenance (non-decrease) of environmental quality, appears to be a realistic approach. Under this aspect, the present cost approach also reflects (social) welfare aspects in its valuation of environmental degradation. Theoretical considerations, recently presented by Pearce, Markandya, Barbier (1990, especially p.9), seem to support this approach.

Measurement and valuation problems of estimating the consequential damage (welfare losses) caused by environmental degradation are formidable. It is also difficult to associate unequivocally particular pollutants with health and welfare effects (for example, health damage caused by air pollution). One approach proposed to assess damage costs is to measure actual expenditures required for the elimination of the damage (Uno, 1989). Such expenditures could be shown separately in the SEEA as possible deductions under welfare aspects (Leipert, 1989). Another approach is to directly estimate health and welfare losses, including the impairment of recreational functions or aesthetic and ethical aspects of the environment. Some of these losses have been estimated by using the willingness-to-pay approach as an approximation of individual ("revealed") preferences or by other methods of contingent valuation (see OECD, 1989). Once comprehensive estimates of the value of damages become available, research projects could be undertaken to associate them with the polluting sectors. In this case, separate accounts should be established, which would allow a comparison of the actual and hypothetical avoidance cost on one hand and of the actual and imputed damage costs on the other. These comparisons would facilitate macroeconomic cost-benefit analyses, as proposed for example by Peskin (1989). Such additional accounts would permit further modifications of the components of final demand for the derivation of welfare-oriented measures (Bartelmus, 1987). In the SNA Handbook on Integrated Environmental and Economic Accounting, an approach will be discussed which could be derived from the cost-oriented measures of the SEEA by extending not only the asset boundary, but also the production boundary. This implies the introduction of the concept of environmental services "produced" by nature (see Peskin, 1989; Schäfer and Stahmer, 1989; and Stahmer, 1990).

6.7 APPLICATION OF THE FRAMEWORK: A DESK STUDY OF COUNTRY X

The environmentally modified concepts developed in the framework should stimulate alternative economic analyses and policies, based on an integrated assessment of environment-economy relationships. One aspect of such analysis is to focus on income available for spending on final consumption and new investments. Due to the consideration of environmental costs of production, environmentally adjusted income would generally be lower than income derived in traditional accounting. This *welfare* aspect of environmental accounting has received most of the attention in environmental studies of income and expenditure.

On the other hand, production cost and tangible asset and resource requirements of production reflect a *productivity* aspect of economic performance and environmental-economic analysis. Environmental accounting may result in values of value added generated and tangible assets used in each sector, that are different from the values of income and capital in traditional accounting. The reasons are the inclusion of cost due to environmental uses and of non-produced natural assets in broader concepts of cost and capital, respectively. Changed relations between value added and economic assets used in production might well lead to considerable re-assessment of the rentability and productivity of economic sectors from an environmental (accounting) point of view.

(a) *Economic and Environmental Features of Country X*

This analysis is done on the basis of an illustrative database, developed for the clarification of the above-described environmental accounting concept and procedures. Only part of this database is reflected in Tables 1 to 5 above. The data describe the economic and environmental features of a realistic, but fictitious country "X" and are thus to a large extent fictive. There is however a basic core of data which was taken from the national accounts of an existing country. These core data include GDP by activity and expenditure categories, compensation of employees, indirect taxes (net of subsidies), operating surplus, output, intermediate consumption, capital formation and the consumption of fixed capita

All other national accounts data, included in the framework, are elaborated on the basis of assumptions about the type of country, the circumstances under which traditional GDP is being generated in production, the effects of production on the environment, and environmental protection and response carried out by government, enterprises and individuals. These assumptions permit the break-down of aggregate economic data which are part of conventional national accounts, but which could not be compiled from original data sets. Further assumptions about the environmental conditions of the country were made to obtain environmental data for calculating the environmentally adjusted concepts of income and expenditure.

The economic and environmental features of the fictive country X, as reflected in the data and assumptions, are described in the following. The country is a developing country with oil resources, agricultural production, exploitation of timber resources, and fishing activities in rivers, lakes and ocean.

(i) *Tangible Wealth:* Fixed assets consist of buildings, machinery and equipment, roads and other public structures, and also of livestock for breeding, draught and dairy, trees in orchards and grapevines in vineyards. As regards land, the assumption is that it is used mainly in agriculture for the cultivation of crops and rearing of livestock, in other services for dwellings and office buildings, and that it is owned as infrastructure by government (roads, dams and other structures).

(ii) *Environmental Protection:* Environmental protection activities are carried out in the country by all sectors. They are concentrated however, in three sectors which are selling protection services: (a) "other services" which provide private waste disposal services, environmental consulting and recycling: (b) government, which provides sanitation services, and (c) trade and transport, which transports wastes to dumping areas and treatment and recycling plants. The environmental protection (sanitation) services offered by the government are sold to a very limited extent, the rest is assumed to be used by the government itself. The value of the cleaning activity is assumed to be equal to the cost (5.0). Households also purchase environmental protection services. These purchases (8.8) are presented as expenditure in the column of "household final consumption" in Table 5.3.

(iii) *Mining Exploitation:* The value of the mineral deposits of the country consists in particular of the value of oil reserves (opening stock in Table 6.5: 261.9). New deposits of oil were found as a result of exploration activities. This is presented in Table 6.5 as "Other volume change" (27.8), exceeding the amount of depletion (−8.9 + 0.9).

(iv) *Natural Growth:* The country has an important agricultural sector, a fishing industry which operates in rivers, lakes and the ocean, and timber tracts where wood is cut and replanted in a controlled exploitation activity. There are also minor wood collection activities in rural areas which are not controlled by any permits.

(v) *Natural Disasters:* During the period of accounting, the country suffered from a major earthquake which destroyed some of its infrastructure, particularly roads owned by the government, machinery and equipment in the manufacturing sector, and dwellings and other buildings that are recorded as capital of the other service sector. The total value of destruction is included in Table 6.5 ("Volume change by natural or multiple causes:"− 25.3).

(vi) *Pollution:* Pollution effects as a result of economic activities in the country are recorded in Table 5.3 in the row corresponding to qualitative degradation of land, water and air by residuals. In the case of air and water pollution, it is assumed that not only the domestic air and water were affected, but also those of neighbouring Countries (−4.7). Private households also cause pollution, which is assumed to consist of the effects of accumulated and illegally discharged wastes. The cost of the pollution (15.6) is reflected in the intersection of qualitative degradation by residuals and the household consumption.

(viii) *Conversion of Tropical Forest of Commercial Use:* Tropical rain forests are being converted to a limited extent to land for agriculture, urbanization and industrial development. This is recorded as change of land use in Table 6.5 (±3.4).

(b) Comparative Analysis of the Economic Conditions of Country X, Based on National and Environmental Accounts

In Tables 5.7 and 5.8 we compare aggregates and indicators from traditional accounting with corresponding ones in environmental accounting. NDP is the main concept of national accounting and EDP is used as the environmental accounting alternative.

Table 6.7: Analytical Measures in Traditional and Environmental Accounting: Resources and Uses

	Based on:		
	NDP	EDP	Percentage Difference
Macro-aggregates	*(1)*	*(2)*	*(2)-(1)/NDP*
Income/expenditure	267.1	185.1	−31
Final consumption	217.5	212.5	−2
% of final domestic uses	81	112	
Capital formation (accumulation) net	50.4	−23.5	−28
% of final domestic uses	19	−12	
Exports	73.7	69.0	−2
Minus: imports	74.5	72.9	−1

(i) *Income and Expenditure:* The income and expenditure analysis is done on the basis of the national accounts identity between income on one hand and domestic expenditure (final consumption, investment) plus exports minus imports on the other. The income and expenditure aggregates, based on Tables 2 and 3, are presented in Table 6.7.

In the case of traditional national accounting (column 1 of Table 6.7) the income concept is NDP, and the expenditure concepts are final consumption, net capital formation and exports minus imports. In the case of environmental accounting (column 2), NDP is replaced by EDP, consumption by environmentally adjusted consumption and net capital formation by net capital accumulation. Environmentally adjusted consumption is derived from final consumption by deducting the improvement in the environment which is assumed to be equal to the government's net (accounting for clean-up of its own pollution) expenditure for environmental protection (5.0). Environmentally adjusted capital accumulation (−23.5) is arrived at by deducting from net capital formation (50.4) total environmental uses for all (produced and non-produced) asset categories (73.9).

The two sets of aggregate data present a very different picture of the economic situation of the country. Income is reduced drastically between NDP and EDP from 267.1 to 185.1 which represents a 31 per cent reduction. Most of this reduction is caused by the modifications of the concept of capital formation to obtain a new concept of capital accumulation. Net capital formation changes from being positive (50.4) to a negative net capital accumulation of −23.5, which constitutes a reduction of income of 28 per cent. The remaining 3 per cent of the total reduction of GDP are explained by the difference between final consumption and environmentally adjusted consumption (−2 per cent) and the decrease of exports minus imports (−1 per cent) (see column 3, Table 6.7).

According to conventional national accounting, the country's domestic expenditure presents a healthy picture of capital formation of 19 per cent of total expenditure. Environmental accounting indicates, however, that capital accumulation has been negative. The main factor explaining this result is the inclusion of non-produced assets into the asset boundary: the depletion of natural assets reduces the capital formation by a value of 18.2; a further reduction (−55.7, see Table 6.3) results from the degradation of land, water and air.

(ii) *Income, Output and Capital:* Production-related changes in environmental accounting, as compared to traditional national accounting, are elaborated on in the three sections of Table 6.8. In Table 6.8 we use figures for specific industries which are not shown in Tables 6.2 and 6.3 above, but which represent disaggregated (by economic sectors) figures of these tables. Section (i) of the table shows that there is a reduction in value added for the economy as a whole with EDP amounting to 69 per cent of NDP. The impact differs from sector to sector, however. The largest reductions are in mining (52 per cent) and agriculture (49 per cent). In manufacturing, the reduction is 22 per cent. Trade and transport also show a reduction of 14 per cent due to the environmental cost of traffic pollution. All other sectors have lower reductions.

Table 6.8: Analytical Measures in Traditional and Environmental Accounting: Income, Output and Capital in Per cent

Analytical Measures	Total	Agriculture	Mining	Manufacturing	Electr. Gas Water	Construction	Trade and Transp.	Other Services	Government Services	Environ- mental Adjustm.
					Industries					
(i)										
EDP as percentage of NDP	69	51	48	78	97	90	86	91	96	
Value added as percentage of output, based on:										
NDP	52	77	44	34	36	45	61	62	61	
EDP	36	39	21	27	34	41	52	57	59	
Value added as percentage of:										
NDP	100	12.1	12.6	16.7	0.8	7.1	26.2	14.5	9.9	
EDP	100	8.9	8.8	18.7	1.1	9.2	32.5	19.1	13.7	– 12
(ii)										
Opening balance sheet:										
Ratio of produced assets/all assets, inclusive of non-produced assets	38	43	13	91	92	95	97	34	29	
Percentage changes between opening and closing balance sheets:										
Net capital formation and volume change due to econ. use (ecol. valuation)										
Produced assets	4	2	0	11	2	3	4	10	3	

Table 6.8 (contd.)

Analytical Measures	Total	Industries								Environmental Adjustm.
		Agriculture	Mining	Manufacturing	Electr. Gas Water	Construction	Trade and Transp.	Other Services	Government Services	
All assets, incl. non-prod. assets	1	- 4	- 2	10	2	3	4	3	1	
Other volume changes in asset, net										
Produced assets	- 2	0	0	- 16	0	0	0	- 5	- 1	
All assets, incl. non-prod. assets	0	0	8	- 14	0	0	0	- 2	- 1	
Revaluation and environ, value discrepancies										
Produced assets	14	15	13	13	15	13	12	15	15	
All assets, incl. non-prod. assets	20	20	13	14	16	13	13	23	24	
Total changes, net										
Produced assets	16	17	13	8	17	15	16	20	17	
All assets, incl. non-prod. asets	21	17	19	10	18	16	16	24	24	

(iii)

	Total	Agriculture	Mining	Manufacturing	Electr. Gas Water	Construction	Trade and Transp.	Other Services	Government Services	
Value added-capital ratios, based on:										
NDP	25	20	73	46	35	21	44	49	6	
EDP	7	4	5	32	31	18	36	15	2	

These differences are also reflected in the ratios of value added over output under NDP and EDP calculations in section (i) of Table 6.8. For the economy as a whole, the ratio falls from 52 per cent to 36 per cent. The largest drop is in agriculture from 77 per cent to 39 per cent, followed by mining (44 per cent to 21 per cent), trade and transport (61 per cent to 52 per cent), and manufacturing (34 per cent to 27 per cent). Other sectors show much lower reductions. Consequently, there are changes in the order of sector contributions to EDP as compared to NDP. Trade and transport is the largest contributor to both EDP and NDP. Manufacturing is the second largest and other services the third largest contributor of NDP. For EDP, however, this order is inverted. The weight of agriculture and mining in the economy is decreased whereas construction and government services increase in importance.

The other element of production cost, the use of economic wealth, is also affected by differences in coverage between traditional national accounting and environmental accounting. As shown in section (ii) of Table 6.8, produced assets, which are the capital element in national accounting, for the economy as a whole amount to only 38 per cent of the total value of capital used, if non-produced assets are taken into account. For individual sectors the differences in coverage of economic wealth used in production is even more pronounced: particularly in the mining sector, produced assets are only 13 per cent of the total value of assets used by economic activities.

Changes in coverage of assets also affect the change over time of economic wealth between opening and closing balance sheets. In section (ii) of Table 6.8, the total changes in assets are broken down into net capital accumulation (including ecologically valued volume change of natural assets due to economic use), other volume change and valuation discrepancies due to market price changes and adjustments of ecological assets to market valuation.

The percentage change attributed to net capital accumulation is higher for produced assets (4 per cent) than for all assets (1 per cent). For other volume changes, however, this relationship is inverted. In the case of the traditional capital concept, other volume changes— due to earthquake damages—cause a reduction of produced assets (−2 per cent), while economic wealth based on the broader concept roughly remains unchanged (0 per cent). This inversion is mainly the result of the inclusion of new finds of subsoil assets (27.8, see Table 6.5) in the latter concept.

Discrepancies in valuation amount to 14 per cent for produced assets, reflecting the average annual inflation in the country which was assumed to be 15 per cent. In contrast to this, the value of all assets is increased by 20 per cent. This is mainly due to the inclusion of the asset of cultivated land with its high price increases (331.0 on an opening stock of 1,366.7, see Table 6.5).

For the individual sectors a basic pattern can be described: total volume changes, defined as the sum of net capital accumulation and other volume changes, are approximately the same between the traditional capital concept and the broader economic wealth concept. There is a marked difference in the case of agriculture: in this sector, the volume of economic wealth increases by 2 per cent (net capital accumulation of 2 per cent plus other volume changes of 0 per cent) when the narrower concept of produced assets is used, while it decreases by 4 per cent (net capital accumulation of –4 per cent plus volume changes of 0 per cent) when volume changes in all assets are taken into account. The latter reduction is the result of the negative effects of land erosion, depletion and pollution (including acid deposition) on natural resources held by agriculture, forestry and fishing.

The combined changes in value added and economic wealth used in economic activities have considerable effects on the productivity or rentability of capital. The different effect on NDP- and EDP-based value added/capital ratios are presented in the last part (iii) of the table. For the economy as a whole, the ratio between value added and capital based on national accounting (NDP) is 25 per cent; this is reduced to 7 per cent if based on environmental accounting (EDP). For specific sectors, the differences are even larger. In mining, the value added/capital ratio based on NDP is 73 per cent while for EDP it is only 5 per cent. For agriculture, the ratio is reduced from 20 per cent to 4 percent and for other services from 49 per cent to 15 per cent. These are significant changes in productivity or rentability indicators which might prompt a ressessment of investment policies as far as capital allocation to economic sectors is concerned.

To the extent that environmental cost are also included in (internal) business accounts, new EDP-based measures might also affect micro-economic investment decisions.

End Notes

1. See *e.g.* the Norwegian approach to natural resource accounting (Alfsen, Bye and Lorentsen, 1987) or the more complex (including interactions in the biophysical environment) French "natural patrimony"

accounts (Institute National, 1986).

2. This treatment of natural growth processes in agriculture, forestry and fishing differs from the 1968 SNA recommendations, but may be adopted in the revised SNA.

3. See El Serafy (1989). However, the total depreciation approach is advocated by Harrison (1989) and Repetto (1989).

4. The normal rent of capital refers to the produced assets which have been used for the exploitation of natural assets (*e.g.* trawlers for fishing and drilling instruments).

5. The question of welfare-oriented measures in national accounts is discussed in Drechsler (1976) and United Nations (1977a). The limits of accounting approaches to assets, the sustainability of economic growth, and possibilities of modeling the "feasability" of development programs are discussed in Bartelmus (in preparation).

References

Ahmed, Y.J., El Serafy, S., and Lutz, E. (eds.). *Environmental Accounting for Sustainable Development*, The World Bank, Washington, D.C., 1989.

Alfsen, K.H., Bye T., and Lorentsen, L., *Natural Resource Accounting Analysis, The Norwegian Experience, 1978–1986*, Central Bureau of Statistics of Norway, Oslo, 1987.

Ayes, R.U., *Resources, Environment and Economics*, New York, 1978.

Ayres, R. U. and Kneese, A. V., Production, Consumption and Externalities, *American Economic Review,* Vol, 59, No.3, 282–297, June 1969.

Bartelmus, P., Accounting for Sustainable Development, United Nations, Department of International Economic and Social Affairs, Working Paper No.8, New York, 1987.

————, *Accounting for sustainable Growth and Development: Structural Change and Economic Dynamics* (in preparation).

Blades, D.W., Measuring Pollution within the Framework of the National Accounts, in Ahmad, Y.J., El Serafy, S., and Lutz, E., 26–31, 1989.

Daly, H.E., Sustainable Development: From Concepts and Theory towards Operational Principles, *Populations and Development Review* (in preparation).

————, Sustainable Development: From Concepts and Theory towards Operational Principles, (in preparation).

Drechsler, L., Problems of Recording Environmental Phenomena in National Accounting Aggregates, *Review of Income and Wealth*, 22(3), 239–252. 1976.

El Serafy, S., The Proper Calculation of Income from Depletable Natural Resources, in Ahmad, Y.J., El Serafy, S., and Lutz, E., 10–18, 1989.

Harrison, A., Environmental Issues and the SNA, *Review of Income and Wealth,* 35 (4),377–388, December 1989.

Hueting, R., *New Scarcity and Economic Growth: More Welfare Through Less Production*: 1980.

Hueting, R. and Leipert, C., Economic Growth, National Income and the Blocked Choices for the Environment, Wissenschaftszentrum Berlin fur Sozialforschung (discussion paper), Berlin, 1987.

Institut National de la Statistique et des Etudes Economiques *Les Comptes du Patrimoine Naturel*, Les collections De I'INSEE, Ser. C, 137/138, INSEE, Paris, 1986.

————, *Les Comptes Satellites de l'Environnement, Methods et Resultants,* Les collections De I'INSEE, Ser. C 130, INSEE, Paris, 1986a.

Israd, W., Some Notes on the Linkage of Ecologic and Economic Systems, *Regional Science Association Papers*, 22, 85–96, 1969.

Leipert, C. National Income and Economic Growth: The Conceptual Side of Defensive Expenditures, *Journal of Economic Issues,* 23(3), 843–856, 1989.

Leontief, W., National Income, Economic Structure and Environmental Externalities, in Moss, M. (ed.), *The Measurement of Economic and Social Performance*, Studies in Income and Wealth, Vol. 38, NBER New York, London, 565–578. 1973.

Lemaire. M. Satellite Accounting: A Solution for Analysis in Social Fields, *Review of Income and Wealth*, 33(3), 305–325, 1987.

Lutz, E. and El Serafy, S., Recent Developments and Future Work, in Ahmad, Y.J., El Serafy, S., and Lutz, E., 88–91, 1989.

Martinez, A.R., Ion, D.C., De Sorcy, G.J., Dekker, H., and Smith, S., Classification and Nomenclature System for Petroleum and Petroleum Reserves, Twelfth World Petroleum Congress, Houston, 1987.

OECD Environmental Policy Benefits: Monetary Valuation, study prepared by D.W. Pearce and A. Markandya, Paris, 1989.

Olsen, M. The Treatment of Externalities in National Income Statistics, in Wingo, L. and Evans, A. (eds.) *Public Economics and the Quality of Life*, Baltimore, Md., 1977.

Pearce, D.W., Markandya, A., and Barbier, E., *Blueprint for a Green Economy*, London, 1989.

————, *Sustainable Development: Economy and Environment in the Third World,* London, 1990.

Peskin, H. M., A Proposed Environmental Accounts Framework, in Ahmad, Y.J., El Serafy, S., and Lutz, E., 65–78, 1989.

Pezzey, J., Economic Analysis of Sustainable Growth and Sustainable Development, World Bank, Environment Department Working Paper No. 15, Washington, D.C., 1989.

Repetto, R. and others, Wasting Assets: Natural Resources in the National Income Accounts, World Resources Institute, June 1989.

Schäfer, D. and Stahmer, C., Input-output Model for the Analysis of Environmental Protection Activities, *Economic Systems Research,* 1(2), 203–228, 1989.

————,Conceptual Considerations on Satellite Systems. *Review of Income and Wealth*, 36(2), 167–176, 1990.

Stahmer, C. Cost-Oriented and Welfare-Oriented Measurement in Environmental Accounting, paper presented at the Fifth Karlsruhe Seminar on Models and Measurement of Welfare and Inequality, August 12–19, 1990.

Teillet, P.A Concept of Satellite Accounts in the Revised System of National Accounts, *Review of Income and Wealth*, 34 (4), 411–439, 1988.

United Nations, *A System of National Accounts*, United Nations publication, No. E.69. XVII.3,1968.

―――――, *Draft Guidelines for Statistics on Materials/Energy Balances*, United Nations Publication No. E/CN.3/492,29 March, 1976.

―――――, *Provisional International Guidelines on the National and Sectoral Balance-Sheet and Reconciliation Accounts of the System of National Accounts*, United Nations publication, No. E.77. XVII.10, 1977.

―――――, *The Feasibility of Welfare-Oriented Measures to Supplement the National Accounts and Balances: A Technical Report*, United Nations Publication Series, F, No. 22, New York, 1977.

―――――, *Revised System of National Accounts, Preliminary Draft Chapter, Provisional future* ST/ESA/STAT/SER.F/2/Rev.4, February 1990.

―――――, *International Standard Industrial Classification of all Economic Activities* (ISIC), United Nations Publication, Series M, No.4, Rev. 3, New York, 1990a.

―――――, *Concepts and Methods of Environment Statistis, Statistics of Natural Environment―A Technical Report*, United Nations Publication, in preparation.

―――――, *Central Product Classification* (CPC), United Nations Publication, in preparation.

Uno, K., Economic Growth and Environmental Change in Japan–Net National Welfare and Beyond, in Archibugi, F. and Nijkamp, P. (eds.), *Economy and Ecology: Towards Sustainable Development*, 307–332, Dordrecht, 1989.

Ward, M., *Accounting for the Depletion of Natural Resources in the National Accounts of Developing Countries*, OECD Development Centre Publication, Paris, 1982.

World Commission on Environment and Development, *Our Common Future*, Oxford University Press, Oxford, 1987.

CHAPTER

7

Formal Models and Practical Measurement for Greening the Accounts

Kirk Hamilton

7.1 INTRODUCTION

Over the past 25 years the breadth of environmental and natural resource policy making has expanded from dealing with pollution incidents, such as the grounding of an oil tanker, to grappling with the new complexities of achieving 'sustainable development' and protecting the global commons. Governments everywhere have committed themselves to these new goals, whether as a product of the United Nations Conference on Environment and Development in 1992, or in response to the report of the Brundtland Commission in 1987. Policy makers need new measures of progress.

There has been corresponding innovation in information systems to guide resource and environmental policies, from the collection of physical and economic data to the development of conceptual frameworks. The OECD's 'Pressure-State-Response' framework, for instance, provides the means to interrelate complex information concerning human activities and the environment. Environmental performance indicators have taken on a key role in many countries. And natural resource and environmental accounts, with their tight coupling to *economic* accounts and indicators, promise to provide policy makers with measures of progress towards environmentally sustainable development.

The development of environmental and natural resource economics has highlighted the critical role that policy failure and market failure play in the degradation of the environment. This has put the focus squarely on policy, where it belongs. Policy failures come in many forms, from inadequate property rights regimes, to underpricing of natural resources, to subsidies on energy, fertilizers and pesticides that lead to negative impacts on the environment. Market failure exists wherever economic activities impose costs on others, in the form of pollutants carried downwind or downstream for instance, without any mechanisms for remediation.

Market failure is the prime reason that many analysts question the policy signals provided by our traditional economic indicators, in particular the Gross National Product (GNP). As codified in the UN System of National Accounts (SNA) (United Nations, 1993a), GNP measures the sum total of economic production on the basis of transactions in the marketplace. As a result, GNP masks the depletion of natural resources and presents an incomplete picture of the costs imposed by the polluting byproducts of economic activity. This has led many people to conclude, as in the case of Repetto *et al.* (1989, p. 3), that:

> This difference in the treatment of natural resources and other tangible assets [in the existing national accounts] reinforces the false dichotomy between the economy and 'the environment' that leads policy makers to ignore or destroy the latter in the name of economic development.

The new emphasis that governments have placed on sustainable development is another source of criticisms of the traditional national accounts. Measures such as Net National Product (NNP) and National Income account are used only for the depreciation of produced assets, ignoring the value of depletion of natural resources and degradation of the environment. They cannot serve, therefore, as guides for policies aimed at achieving sustainable development. 'Greener' aggregates, it is hoped, can.

In addition to these criticisms of traditional national accounting aggregates, natural resource and environmental accounting has many other antecedents. The experience of the 'oil crisis' in the 1970s led to a concern with the physical scarcity of natural resources, and the construction of natural resource accounts detailing the changing stocks and flows of physical quantities of resources was the result (Alfsen, 1993). Worries about the toxic effects of pollutants led to increasing interest in understanding the pathways that particular materials take through the economic system, with the development of material balance accounts being one of the responses (Ayres and Kneese, 1969). And the desire to analyse the connection between economic activity and pollutant flows produced models linking input–output tables to accounts of pollution emissions (Leontief, 1970; Victor, 1972; see also

Chapter 7). All of these roots are still evident in what is broadly termed resource and environmental accounting.

The United Nations has attempted to bring some order to this field with the publication of interim guidelines on an integrated System of Environmental and Economic Accounts (SEEA) (United Nations, 1993b). This system aims to provide a common framework within which greener national accounting aggregates, natural resource accounts and pollution flow accounts have their appointed places.

For developing countries, the adjustments to standard national accounting aggregates that result from resource and environmental accounting are sizable, as will be seen in the empirical results below. This is obviously true for the most resource-dependent economies, but it is also likely to be of growing importance for those countries that are rapidly industrialising and urbanising; for these countries the growth in damages from pollution emissions, in terms of human health in particular, is of mounting concern.

This chapter aims to outline the variety of environmental accounts that countries and institutions are constructing, to explain the theoretical basis for environmental accounts, and to summarise what theory tells us about the mechanics of adjusting the national accounts to reflect environmental change. To show that these adjustments matter, recent empirical work from the World Bank will be presented. Finally, there will be a brief discussion of the types of policy questions raised by greening the accounts.

7.2 VARIETIES OF ENVIRONMENTAL ACCOUNTS

As noted in Section 1, there are several different flavours or approaches to environmental accounting. While many taxonomies are possible, the breakdown that most closely matches what countries are actually doing is the following:

- *Adjusted national accounting aggregates,* such as Eco-Domestic Product (EDP) or expanded measures of saving and wealth.
- *Natural resource accounts,* where the emphasis is on balance sheet items (the opening and closing stocks of various natural resources) and the flows that add to and subtract from the balance sheet position; these accounts are in quantities and (possibly) values, and may or may not be linked to the SNA through the National Balance Sheet Accounts.
- *Resource and pollutant flow accounts* that embody consider-

able sectoral detail and often are explicitly linked to the input-output accounts, a part of the SNA.

- *Environmental expenditure accounts*, which represents a breakout of existing figures in the SNA.

These are described in more detail below.

7.2.1 Adjusted National Accounting Aggregates

Hicks (1946, p. 173) defined income as:

> ... the maximum amount which can be spent during a period if there is to be an expectation of maintaining intact the capital value of prospective returns ...; it equals consumption plus capital accumulation.

The notion of Hicksian income underpins most economic thinking about income and is embodied in the System of National Accounts. The obvious extension of Hicks into the realm of environmental accounting is to include in 'capital accumulation' the changes in value of a wider range of economic assets, including in particular those environmental assets that are scarce and valued by people. Such a definition will exclude some things, such as the value that the environment may yield for other species, but can encompass much of what humans value in the environment.

This extended Hicksian framework provides the basis for altering the standard SNA aggregates. The key steps in defining and measuring adjusted national accounting aggregates are as follows:

- The starting point is simply the balance sheet account from the 1993 revision of the SNA. In addition to the value of produced assets (buildings, machinery and equipment, infrastructure), this account includes a variety of tangible non-produced assets including land, subsoil assets, non-cultivated biological resources, and water. The key stipulations with regard to these assets are that they occur in nature, and that they have ownership regimes that are enforced and may be transferred (this limits, for instance, which water resources may be valued in the balance sheet accounts).
- Changes in wealth over the accounting period serve to define an adjusted savings measure, termed 'genuine' saving in recent literature.[1] This is measured as gross saving, less depreciation of produced assets, less the depletion of natural resources, less damage to tangible assets (both produced and non-

produced, for example, acid rain damages to soil fertility), less
damage to human health. The latter should be valued on the
basis of willingness-to-pay, and should exclude any productivity losses that may already be reflected in the national accounts.

- Adjusted income or EDP is measured as the sum of consumption (from the standard SNA) and genuine saving. This is the
Hicksian 'consumption plus capital accumulation'.

7.2.2   Natural Resource Accounts

As previously noted, natural resource accounts generally have a
balance sheet flavour, with their emphasis on opening and closing
stocks, in quantities and values, of natural resources including both
commercial natural resources and non-commercial or environmental
resources. As such, resource accounts form the basis of the expanded
national balance sheet accounts in the revised SNA. The principal
policy and analytical uses of these accounts include:

- Measuring physical scarcity.
- Providing information for resource management.
- Productivity measurement; valuing the natural assets upon which
the resource sectors depend, can help to refine measures of
national productivity.
- Valuing depletion.
- Measuring the economic impact of pollution on living natural
resources such as forests and fisheries.

Natural resource accounts, and their counterparts in the national
balance sheet accounts, can therefore have wide use with regard to
resource management policies and broader environmental policies.

7.2.3   Resource and Pollutant Flow Accounts

Resource and pollutant flow accounts are generally conceived as
physical extensions to the (monetary) input-output (I–O) accounts.
For each production and final demand sector in the I–O tables, these
accounts associate a physical flow of natural resources, typically as
inputs such as energy to production processes, and a physical flow of
wastes and emissions in the form of SO_2, NO_x, BOD, etc. With links
to the I–O tables these accounts lend themselves naturally to policy
modelling. Examples of policy uses include:

- Measuring the incidence of environmental regulations and taxes.
- Estimating emission tax rates in conjunction with computable general equilibrium models.
- Measuring the efficiency of resource use.
- Macro-modelling; tying resource and pollutant flow accounts to the standard macroeconomic models governments use for projections permits the reporting of environmental effects (in terms of resource throughput and pollution emissions) as a standard component of the output from such models.
- Dispersion and impact models; the estimation of pollution emissions is a required input for 'downstream' models of dispersion and impact.

Of the *physical* accounts under consideration, resource and pollutant flow accounts clearly have links to the widest variety of policy issues. Construction of extended physical flow accounts has been pursued with particular vigour in Europe, under the heading of 'National Accounting Matrices with Environmental Accounts' (NAMEA) (see De Haan *et al.*, 1993).

7.2.4 Environmental Expenditure Accounts

Environmental expenditure accounts generally consist of detailed data on capital and operating expenditures by economic sectors, for the protection and enhancement of the environment. The accounts may or may not include detail on the type of pollutant controlled, or the environmental medium being protected. The prospective uses of these accounts are fairly straightforward:

- Measurement of the total economic burden of environmental protection.
- Measurement of the sectoral distribution of abatement and protection costs.
- Measurement of unit abatement costs; if the survey vehicles used to collect data on environmental expenditures also collect data on the amount of abatement achieved, it is possible to estimate average unit abatement costs.

It should be noted, however, that measuring environmental expenditures is a subject fraught with definitional and measurement problems. To give just one example, manufacturers now often introduce new production technologies that jointly increase productivity and decrease emissions. In such a case, there is no meaningful way to establish what is the cost of protecting the environment.

The exposition in the balance of this chapter will focus on the construction of 'greener' environmental accounting aggregates. This topic presents both the greatest challenge to environmental analysts and the greatest opportunity for impacts on policy.

7.3 THE THEORETICAL BASIS FOR ENVIRONMENTAL ACCOUNTS

Weitzman (1976) posed a fundamental question about economic accounting: why, if the economic goal is to maximise consumption, do we measure national income as the sum of consumption and investment? One way to answer this question is to note that the sum of consumption and investment is what one would choose to maximise at each point in time if the economic goal were to maximise the *present value* of consumption along the optimal path for the economy. The Weitzman paper suggests that there is a profound linkage between national income accounting and growth theory.[2]

The notion of Hicksian income was introduced above, and a straightforward extension of this concept leads to a generalised definition of national income. It is traditional in growth theory to define total wealth to be the present value of consumption along the optimal path; in certain simple growth models this can be shown to be equal to the current value of the capital stock. If instead we define total wealth to equal the present value of *welfare* along the optimal path, then extended Hicksian income can be defined as the maximum amount of produced output that could be consumed, while leaving total wealth instantaneously unchanged.

These ideas can be made more concrete by presenting a fairly general model of an economy with natural resource stocks, and stocks of pollution associated with levels of production activity.

Assume a simple closed economy with a single resource used as an input to the production of a composite good that may be consumed, invested in produced assets or human capital, or used to abate pollution, so that $F(K, R, N) = C + \dot{K} + a + m$, where R is resource use, a is pollution abatement expenditures, N is human capital, and m is investment in human capital (current education expenditures). Function $q(m)$ transforms education expenditures into human capital that does not depreciate (it can be considered to be a form of disembodied knowledge), so that $N = q(m)$.[3] Labour is fixed and is therefore factored out of the production function.

Pollution emissions are a function of production and abatement, $e = e(F, a)$, and pollutants accumulate in a stock X such that $\dot{X} = e - d(X)$ where d is the quantity of natural dissipation of the pollution stock. The flow of environmental services, B, is negatively related to the size of the pollution stock, so that $B = a(X)$, $a_X < 0$. Resource stocks, S, grow by an amount g and are depleted by extraction, R, so that $\dot{S} = -R + g(S)$, and resources are assumed to be costless to produce. The utility of consumers is assumed to be a function of consumption and environmental services, $U = U(C, B)$. There is a fixed pure rate of time preference r.

Wealth, V, is defined to be the present value of utility on the optimal path. It is assumed that a social planner wishes to maximise wealth as follows:

$$\max_{C} \int_{t}^{\infty} U(C, B)e^{-rs}ds$$

subject to:

$$\dot{K} = F - C - a - m$$
$$\dot{X} = e = d$$
$$\dot{S} = -R + g$$
$$\dot{N} = q(M)$$

The current value Hamiltonian function, which is maximised at each point in time, is given by:

$$H = U + \gamma_K \dot{K} + \gamma_X \dot{X} + \gamma_S \dot{S} + \gamma_N \dot{N}. \qquad (7.1)$$

where γ_K, γ_X, γ_S and γ_N are respectively the shadow prices in utils of capital, pollution, resources and human capital. Deriving the static first-order conditions for a maximum, the Hamiltonian function may be written as:

$$H = U(C, B) + U_C (\dot{K} - (1 - be_F)F_R (R - g) - b(e - d) + q/q').$$

Note that b is the marginal cost of pollution abatement. It is shown in Hamilton (1996) that this is precisely equal to the marginal social cost of pollution emissions, and that this in turn is equal to the level of a tax (the Pigovian tax required to maximise welfare) on emissions. These equalities hold because the economy is at the optimum. The term be_F is the effective tax rate on production as a result of the emissions tax. Therefore, although we have started with an optimal growth problem, the prices that result are those that would prevail in a competitive economy with a Pigovian tax on pollution. Note as well that $1/q'$ is the marginal cost of creating a unit of human

Since $\dot{S} = -R + g$, $\dot{X} = e - d$ and $\dot{N} = q$, the parenthesised expression in the second term of this expression is equal to the change in the real value of assets in this simple economy, where human capital is valued at its marginal creation cost, pollution stocks are valued at marginal abatement costs and natural resources at the resource rental rate, F_R, net of the effective tax rate on production associated with pollution emissions. This expression serves to define genuine saving, G:

$$G = \dot{K} - (1 - be_F) \, F_R \, (R - g) - b(e - d) + q/q'. \tag{7.2}$$

For non-living resources the term in growth, g, can be dropped from expression (7.2), while for cumulative pollutants the term in dissipation, d, can be discarded.

Genuine saving therefore consists of investment in produced assets and human capital, less the value of depletion of natural resources and the value of accumulation of pollutants. It is straightforward to show that:

$$U_C G = \dot{W} = rW - \dot{U} \tag{7.3}$$

Expression (7.3) entails the following property: measuring negative genuine saving at a point in time implies that future utility is less than current utility over some period of time on the optimal path. Negative genuine saving therefore serves as an indicator of non-sustainability by Pezzey's (1989) definition (this will be discussed below).

This expression also implies that Hicksian income, the maximum amount of produced output that could be consumed while leaving total wealth instantaneously constant, is given by:

$$NNP = C + \dot{K}. - (1 - be_F)F_R \, (R - g) - b(e - d) + q/q' \tag{7.4}$$

The treatment of current education expenditures and pollution abatement expenditures requires more elaboration. Hamilton (1994) essentially argues that current education expenditures are not consumption, and therefore should be included in saving. Defining net marginal resource rents as $n \equiv (1 - be_F) \, F_R$, NNP be can be defined as:

$$NNP = GNP - a - m - n(R - g) - b(e - d) + q/q'$$

$$= GNP - a - n(R - g) - b(e - d) + \left(\frac{1/q'}{m/q} - 1 \right) m. \tag{7.5}$$

where $1/q'$ is the marginal cost of creating a unit of human capital and m/q is the average cost. Assuming increasing marginal education costs, expression (7.5) suggests that the value of investments in human capital should be greater than current education expenditures; these current expenditures can therefore serve as a lower-bound estimate of the investment in human capital.

7.4   WHAT THEORY TELLS US

The growth model just presented is clearly far too aggregate and simple to be used in real-world problems. Its value, when combined with the Hicksian framework, is to give clear guidance on two critical questions for green national accountants: what should be subtracted (and what added) in constructing measures of income and saving? And how should these subtractions and additions be valued?

While many of the lessons from theory highlighted below are derived from the model just presented, others arise in the recent literature on environmental accounting. The lessons from theory fall under three headings: natural resources, pollution, and other issues.

7.4.1 Natural Resources

For living natural resources, expression (7.4) gives clear guidance: net depletion (harvest minus growth) should be deducted from NNP. If growth exceeds harvest, then there is an argument for actually increasing NNP to reflect the growth of this stock. However, it is important to count only growth in the economic portion of the total living stock (*i.e.* that portion with positive economic rents).[4] The basis of valuation is the unit scarcity rent, attenuated by the share of pollution damages in GNP.

Deforestation is not covered by expression (7.4). This represents a change in land value associated with a change in land use. Hamilton *et al.* (1998) show that *excess* deforestation should be deducted from NNP; this excess is determined by the number of hectares cleared for which the total economic value (TEV) of standing trees exceeds the TEV of the new land use (typically farmland). TEV for fcrested land includes both local and global willingness to pay for the services provided by standing forest, sustainable off-take of timber and non-timber products, production externalities (the value of upland forests to lowland agriculture, for example), and any carbon sequestration benefits.

For non-living resources, this expression suggests that depletion should be measured as the quantity of resource extracted, valued at the unit scarcity rent (*i.e.* the marginal rent). Because very strong assumptions concerning the optimality of the resource extraction program are required to support this valuation, and since marginal rents are typically not easily measured, the literature generally favours two valuation methods which arguably span a reasonable value for depletion: the simple present-value approach (El Serafy, 1989), which values depletion as the present value of current total resource rent (the reserves to production ratio defines the time period for the present value), or the total-rent approach (as defined, for example, in Repetto *et al.*, 1989) which values depletion as quantity extracted times the average unit resource rent.[5]

7.4.2 Pollution

Expression (7.5) says that pollution abatement expenditures, *a*, are essentially intermediate in character and therefore any *final* abatement expenditures should be deducted in measuring Hicksian income and genuine saving. In practice, most current abatement expenditures are already in fact intermediate inputs in standard national accounting.

Two further issues arise in accounting for pollution: how to deal with abatement expenditures and how to value damages. Abatement expenditures were discussed at the end of the preceding section; any final expenditures on pollution abatement should be deducted from GNP, thereby treating them as intermediate in nature.

On valuing pollution damages, there is a strand of the literature (for example, the UN SEEA (United Nations, 1993)) which argues for a 'maintenance cost' approach. This is the cost of reducing pollution emissions to a given level as defined by a legislative standard or a scientifically determined threshold. Expressions (7.4) and (7.5) suggest that there is no theoretical basis for this approach. Moreover, maintenance costs may seriously understate the required adjustment to GNP, given that in an over-polluted economy, marginal pollution damages will be greater that marginal abatement costs. Maintenance cost calculations do answer a key policy question, however, namely what is the likely cost of meeting a given target for pollution reduction.

Expression (7.4) suggests that the deduction from NNP for pollution emissions should equal marginal damages times the net accumulation of the pollutant. Deriving the efficient path for marginal pollution damages in the model of the preceding section yields the following result:

$$b = \frac{\dot{b} - U_X/U_C}{(1 - be_F) F_K + d'}. \qquad (7.6)$$

It is simplest to interpret this expression in the steady state (as $\dot{b}$ ® 0). Then it can be seen that marginal pollution damages represent the present value of willingness to pay to avoid the pollution over the lifetime of the pollutant in the stock. The numerator is the willingness to pay expression (recall that the marginal utility of pollution is negative), while the denominator is the interest rate (the marginal product of capital adjusted for pollution damages) plus the marginal rate of dissipation of the pollution stock. For long-lived pollutants the marginal rate of dissipation will be very small; it is about 0.5 per cent per annum for CO_2, for example.

This is a good model of stock pollution, or even pollution with cumulative effects, in which case all the terms in d fall out of the preceding expressions. For important pollutants with relatively short atmospheric lifetimes, such as PMO or acid emissions, the accounting should deal with the effects of these pollutants on other economic assets. Empirically, it turns out that the most important of these assets is human health. For health assets, the change in the stock is measured by the number of excess deaths or the number of symptom days, while the unit value of these changes is measured by marginal willingness to pay. Marginal willingness to pay in turn is measured by the 'value of a statistical life'[6] in the case of excess deaths, or the willingness to pay to avoid a marginal symptom day in the case of morbidity.

Avoiding double-counting is an important issue in pollution valuation. For example, many studies of the economic value of pollution damage measure lost productivity as a result of pollution-induced illness as an important component of the damage, which makes perfect sense. But in adjusting the national accounts, it should be remembered that the accounts already reflect the lost production owing to such illness (GNP is lower than it would otherwise be) and so no further adjustment for lost production should be made. To the extent that studies can measure individuals' willingness to pay to avoid the pain and suffering of pollution-induced illness, independent of any lost wages, this can be used in adjusting the accounts as a measure of the change in a notional stock of healthfulness. In general, adjustments to the accounts for pollution health damage should aim to measure welfare losses, rather than productivity losses.

Double-counting can be an issue for other pollution damages as well. For example, acid emissions may damage crops, but this damage is already reflected in lower GNP measures. To the extent that these emissions damage soil fertility over an extended number of years, however, this should be measured in green NNP, since it represents a loss in asset value.

7.4.3   Other Assets

As the formal model of the preceding section makes clear, there are assets other than purely environmental assets that can be brought into an extended set of national accounts. In the case of the formal model, the key non-manufactured, non-environmental asset is human capital, whose creation should arguably be measured in any set of accounts concerned with broader development outcomes. One other example bears mentioning in this context: exogenous resource price changes.

To the extent that a given resource-producing nation is a small player in total world supply of a natural resource, it must operate as a price-taker when it exports its natural resources. The resource price path is therefore exogenous for this nation and this implies that the present value of changes in the resource export value needs to be accounted in any extended measures of income and savings (see Vincent *et al.* (1997) for details). This makes intuitive sense: if the price path for a given resource can be assumed to be upwards independent of any production decisions by the small resource exporter, then there is a type of capital gain (or positive change in the terms of trade) that the exporter will enjoy, and the present value of these gains should be measured in income as a change in asset value. Of course, the opposite implication holds for an assumed declining resource price path.

While there is a clear logic to making such adjustments to the accounts of small resource exporters, the empirical facts suggest there may be little practical significance to this insight. The long-term time trend of most natural resources is, in fact, flat to slightly declining, making any such adjustments to the accounts negligible.

Many other extensions to the accounting system have been explored. Weitzman and Löfgren (1997) argue for an adjustment reflecting the value of exogenous technological change. Aronsson *et al.* (1994) examine the broader accounting issues concerning health.

The question of which assets should be treated in an extended accounting system is, to some extent, a question of the policy issues of concern. If the primary focus is on pollution and natural resources policy, then the sorts of adjustments outlined above probably suffice. If there is a concern with broader development outcomes, however, then there is an argument for dealing not only with natural resources and pollution, but with human capital creation, investments in health, and technological change as well.

7.5 SUSTAINABILITY AND GREEN ACCOUNTING

While it is generally assumed that achieving sustainable development requires a combination of economic, ecological and social sustainability, economists simplify matters by adopting definitions such as that of Pezzey (1989): a development path is sustainable as long as welfare does not decline along the path. As long as welfare is assumed to depend on a wide variety of 'goods' (consumption, environmental amenity, social harmony, and so on), then this definition is capable of capturing important aspects of sustainable development.

One of, the appealing properties of the extended Hicksian measure of income, derived above, is that it leads to a simple test of sustainability: if consumption exceeds Hicksian income, then the economy is not on a sustainable path. Of course, this is just another way of saying that negative genuine saving implies non-sustainability.

Expression (7.3) provides the theoretical justification for genuine saving as a sustainability indicator and, while it is derived for an economy on the optimal development path, a recent paper by Dasgupta and Mäler (forthcoming) shows that this relationship holds for any development path as long as the accounting prices of different assets follow efficient paths. In the real world, we would not expect a point measure of negative genuine saving to imply an unsustainable future, but it is certainly true that a policy mix leading to persistently negative genuine saving is unsustainable.

At first glance, the genuine saving indicator would appear to be limited to questions of *weak* sustainability (*i.e.* the assumption that there are substitutes for all assets). Pearce *et al.* (1989) introduced the important idea of *strong* sustainability, which treats certain natural assets as non-substitutable; a strongly sustainable development path is one which preserves this critical natural capital.

In principle, the genuine saving indicator can deal with strong as well as weak sustainability. Closer examination of expression (7.6) makes this clear. Suppose that there is some critical quantity of pollution stock $\overline{X}$, beyond which there are catastrophic consequences for welfare.[7] The notion of a catastrophic welfare loss would be manifest as extremely large and increasing values of U_X as the critical limit is approached, and a correspondingly large value of marginal willingness to pay, U_X / U_C. This, in turn, implies that the marginal damage of a unit of pollution, b, will also attain elevated values as the limit is approached, which will quickly drive the genuine saving rate to negative values.

As long as there is a well-understood and appropriately steep marginal damage curve for loss of critical assets, the genuine saving measure can capture the phenomenon of strong sustainability. Unfortunately, the nature of many environmental problems is that precise knowledge of the marginal damage curve is not available; this is certainly true for issues such as biodiversity. Under these circumstances, Pearce *et al.* (1996) argue for a precautionary approach, using two indicators: the first is the ratio of the current extent of the critical asset to its threshold value; the second is the genuine saving measure for those elements of nature deemed non-critical and therefore exploitable.

The final point regarding sustainability indicators concerns the role of population growth. The theoretical model presented above is explicitly a fixed population model. If population is growing, then it is possible that genuine saving could be positive, while total wealth per capita is declining. A simple formula makes this clear. Assuming that total wealth is not explicitly a function of population,[8] then for total wealth W and population P it follows that,

$$\Delta\left(\frac{W}{P}\right) = \frac{W}{P}\left(\frac{\Delta W}{W} - \frac{\Delta P}{P}\right) \tag{7.7}$$

where D represents the change in a variable over the accounting period (DW is therefore genuine saving, while DP is the total change in population). If the percentage change in total wealth is less than the percentage growth in population, total wealth per capita will fall.

While expression (7.7) is simple to derive, measuring it presents significant challenges. Population growth rates are widely available, but measures of total wealth are not. Balance sheet accounts are published by relatively few countries, and are limited to the value of produced assets and commercial land (although the 1993 SNA suggests expanding this to include the value of living and non-living natural resources that have an economic value). To be useful in assessing progress towards sustainability, total wealth should include not only produced assets, land and natural resources, but human capital, social capital and intangible assets such as creativity as well. *Expanding the Measure of Wealth* (World Bank, 1997) represents a first step towards total wealth measurement.

7.6 SELECTED EMPIRICAL MEASURES OF GENUINE SAVING

As just noted, genuine saving is a useful indicator of sustainable development. Rather than looking more broadly at other green accounting indicators or types of environmental accounts, therefore, the examples of empirical results will be limited to the savings indicator.

Broad trends in the savings figures can be seen in Table 7.1, which summarises genuine saving for regional and income-level aggregations of countries. These figures are derived from *World Development Indicators* (World Bank, 1999), where a brief description of the methodology is given. More detailed methodological notes appear in Kunte *et al.* (1998).

Comparing low and middle income countries in Table 7.1, the 10.2 per cent difference in genuine savings is largely explained by a 9.2 per cent difference in the gross saving rate. However, depletion is significantly higher in low income countries, at 6.6 per cent of GDP as compared with 4.5 per cent in middle income. Much of this difference in turn is a function of the 1.8 per cent of GDP that net forest depletion represents in low income countries. In high income countries, depletion is only 0.5 per cent of GDP and education expenditures are 2 per cent higher than in low and middle income countries.

Turning to regional comparisons, East Asia and the Pacific exhibits high gross and genuine savings rates, with depletion amounting to 2.1 per cent. As the events of 1997/1998 have made clear, however, high savings rates are not synonymous with financial and macroeconomic stability, however advantageous they may be for rapid wealth accumulation. In Latin America and the Caribbean, the average genuine saving effort is fairly robust, although this masks some individual examples of poor performance, as in the case of Venezuela. Other regions exhibit a weak genuine saving effort, particularly in the oil states of the Middle East and North Africa,[9] and in Sub-Saharan Africa, where depletion is a substantial 7.8 per cent of GDP.

Country-level results for genuine saving in Sub-Saharan Africa in 1997 are presented in Table 7.2. As this table indicates, negative genuine saving is more than a theoretical possibility. It is important to note several issues with regard to these figures. First, as mentioned earlier, a point measure of genuine saving does not necessarily imply that the country in question is fated for an unsustainable development path; it does imply, however, that continuing the current policy mix is unsustainable. Second, it may be perfectly rational for either extremely poor or extremely rich countries to consume wealth in the short run—in the former case to hold off starvation, in the latter because consuming a very small proportion of wealth entails a low loss of welfare over time. Finally, negative genuine savings rates represent an opportunity not taken: resource endowments represent a type of stored development finance, and some countries choose not to benefit from this natural advantage.

Table 7.1: Genuine Saving as Percentage of GDP, 1997

	Gross domestic savings	Consumption of fixed capital	Net domestic savings	Education expenditure	Energy depletion	Mineral depletion	Net forest depletion	Carbon dioxide damage	Genuine domestic savings
World	22.2	11.7	10.5	5.0	1.2	0.1	0.1	0.4	13.6
Low income	17.0	8.0	9.1	3.4	4.2	0.6	1.8	1.2	4.8
Middle income	26.2	9.2	17.0	3.5	3.8	0.5	0.2	1.1	15.0
High income	21.4	12.4	9.0	5.3	0.5	0.0	0.0	0.3	13.5
East Asia & Pacific	38.3	6.9	31.4	2.1	0.9	0.5	0.7	1.7	29.7
Europe and Central Asia	21.4	13.7	7.9	4.2	4.9	0.1	0.0	1.6	5.6
Latin America & Carib.	20.5	8.3	12.2	3.6	2.7	0.7	0.0	0.3	12.1
Middle East and N. Africa	24.1	8.8	15.3	5.2	19.7	0.1	0.0	0.9	-0.3
South Asia	18.2	9.1	9.1	3.8	2.1	0.4	2.0	1.3	7.1
Sub-Saharan Africa	16.8	9.1	7.8	4.5	5.9	1.4	0.5	0.9	3.4

Source: World Development Indicators (World Bank, 1999)

Table 7.2: Genuine Saving and its Components in Sub-Saharan Africa: Percentage of GDP, 1997

	Gross domestic saving	Consumption of fixed capital	Net domestic savings	Education expenditure	Energy depletion	Mineral depletion	Net forest depletion	Carbon dioxide damage	Genuine domestic savings
Angola	27.3	6.0	21.2	2.6	20.7	0.0	0.0	0.4	2.7
Benin	10.8	5.4	5.4	0.0	0.0	0.0	0.0	0.2	5.2
Botswana	44.7	13.3	31.4	6.9	0.0	0.8	0.0	0.3	37.2
Burkina Faso	9.2	4.6	4.6	2.8	0.0	0.0	0.0	0.2	7.1
Burundi	2.6	4.4	−1.8	3.0	0.0	0.0	8.5	0.1	−7.4
Cameroon	20.6	7.5	13.1	2.3	7.4	0.0	0.0	0.3	7.7
C. African Rep.	6.7	5.2	1.5	3.8	0.0	0.0	0.0	0.1	5.1
Chad	1.2	4.6	−3.4	0.0	0.0	0.0	0.0	0.0	−3.5
Congo, Dem. Rep.	9.0	5.0	4.0	0.7	0.0	0.6	0.0	0.2	3.8
Congo, Rep.	34.8	9.2	25.6	4.3	23.9	0.0	0.0	0.4	5.6
Côte d'Ivoire	23.1	7.0	16.0	5.7	1.5	0.0	0.0	0.7	19.6
Eritrea	−17.4	4.1	−21.5	–	0.0	0.0	0.0	–	–
Ethiopia	8.7	–	–	2.9	0.0	0.0	0.0	0.3	–
Gabon	48.3	15.2	33.1	3.1	15.8	0.0	0.0	0.4	20.0
Gambia, The	3.8	12.3	−8.6	3.3	0.0	0.0	0.0	0.3	−5.6
Ghana	9.8	4.3	5.5	2.4	0.0	0.0	0.0	0.4	5.0
Guinea	18.7	6.1	12.6	2.3	0.0	2.5	0.0	0.4	5.0
Guinea-Bissau	5.0	4.8	0.3	1.8	0.0	18.8	0.0	0.2	−4.1
Kenya	11.4	6.7	4.7	5.9	0.0	0.0	0.0	0.5	1.5
Lesotho	−9.8	8.3	−18.1	4.8	0.0	0.0	8.0	0.4	2.1
									−13.3

(contd.)

Table 7.2 (contd.)

Madagascar	3.6	4.9	-1.3	2.3	0.0	0.0	0.2	0.8
Malawi	2.1	6.4	-4.3	3.2	0.0	5.4	0.2	-6.7
Mali	13.6	5.8	7.8	2.8	0.0	0.0	0.1	10.5
Mauritania	8.5	8.6	-0.1	4.9	14.6	0.0	1.7	-11.5
Mauritius	24.1	7.7	16.4	3.1	0.0	0.0	0.0	19.3
Mozambique	13.6	3.6	10.0	3.9	0.0	3.7	0.2	9.9
Namibia	14.2	13.8	0.4	1.7	0.6	0.0	–	1.5
Niger	3.3	4.5	-1.2	1.9	0.0	0.0	0.4	0.2
Nigeria	21.9	2.4	19.5	0.8	0.0	0.0	1.5	-12.0
Rwanda	-7.5	5.6	-13.1	3.2	0.1	0.0	0.2	-10.2
Senegal	13.2	5.3	7.9	4.1	0.4	0.0	0.4	11.1
Sierra Leone	-8.0	5.7	-13.8	2.5	3.6	0.0	0.2	-15.1
South Africa	17.0	13.8	3.2	6.6	1.9	0.1	1.4	4.4
Tanzania	3.4	2.8	0.6	2.9	0.0	0.0	0.2	3.2
Togo	9.8	5.1	4.7	5.3	2.4	0.0	0.3	7.4
Uganda	7.5	5.0	2.5	2.6	0.0	3.4	0.1	1.6
Zambia	9.9	9.9	-0.1	3.8	1.3	0.0	0.4	1.9
Zimbabwe	11.9	6.0	5.9	8.2	9.9	0.4	1.0	2.0

Source: World Development Indicators (World Bank, 1999).

The estimates in Tables 7.1 and 7.2 include the depletion of timber, coal, oil, natural gas, zinc, iron ore, phosphate rock, bauxite, copper, tin, lead, nickel, gold, and silver. Data problems led to the exclusion of diamonds. The only pollutant considered is CO_2, which is valued at $20 per ton of carbon (see Fankhauser, 1995); as suggested by expression (6.6), this represents the present value of global warming damages over the roughly 200-year residence time of a unit of carbon in the atmosphere. Note that as long as it is assumed that other countries have the right not to be damaged by their neighbour's pollution emissions, it is correct to adjust the national accounts of the emitting country. The fact that local pollutants are excluded from Tables 7.1 and 7.2, for lack of data, implies that in some regions of the world the savings estimates are overstated by several per cent of GDP. This is true for air and especially water pollution in South Asia, and for air pollution in East Asia.

The genuine saving estimates in the *World Development Indicators 1999* can be the starting point for a range of analytical investigations. For example, of the six countries with notably negative savings rates in Table 7.2, three (Burundi, Rwanda, and Sierra Leone) are countries facing civil unrest and violence, and another three (Guinea, Mauritania, and Nigeria) are highly dependent on mineral or crude oil resources.

7.7 POLICY ISSUES

Achieving sustainable development will require a range of policy actions, spanning economics, ecology and the social sciences. Greener national accounts can play a role in providing the information base for better policy. The sorts of issues explored below are all concerned with the use of the genuine saving indicator, to the extent that it provides a rough measure of whether an economy is on a sustainable path; the next questions concern the range of policies that can influence this indicator.

It is abundantly clear that monetary and fiscal policies are the biggest levers for boosting savings rates. The first policy issue is, therefore, a classic macroeconomic one: to what extent do monetary and fiscal policies encourage strong domestic savings?

While natural resource exports boost foreign savings and therefore the overall savings effort, the analysis of genuine savings suggests a further question: to what extent do exports of *exhaustible* resources boost the genuine rate of saving?[10] The answer to this lies in netting out the value of resource depletion from gross export values.

More optimal natural resource extraction paths will, other things being equal, boost the value of genuine savings. The policy question for natural resource management is therefore: to what extent can stronger resource policies (royalty regimes, tenure) boost the genuine rate of saving?

Similarly, reducing pollution emissions to socially optimal levels will boost the value of genuine savings. The policy issue with respect to pollution is: to what extent can more optimal pollution control policies increase the rate of genuine saving?

Note that the policy prescriptions for boosting genuine savings should never be to stop extracting resources or emitting pollutants altogether. Rather, pricing resources and pollutants correctly, and enforcing property rights, will lead to efficient levels of exploitation of the environment, reducing incentives to 'high-grade' resources or pollute indiscriminately. Optimal resource and environmental policies will maximise genuine savings, subject to the macroeconomic policy regime in place. However, the sorts of issues raised by Gelb (1988), about the nature and effects of oil windfalls in developing countries, are particularly relevant to the policy issues just raised: without sound macroeconomic policies and prudent allocation of public resources, the effects of reliance upon large resource endowments can be negative for many countries.

7.8 CONCLUSIONS

Still in its development phase, natural resource and environmental accounting is a field with important implications for policies for sustainable development. As governments attempt to match their actions to their rhetoric on achieving sustainability, the importance of environmental accounting will grow.

The development and use of environmental accounting will not be a uniform process across countries. Many developed countries have sophisticated models, that permit the integration of resource and environmental information into macroeconomic analysis. For these countries, the usefulness of adjusting national accounts aggregates may be limited, largely because policy simulations can be carried out directly. The physical natural resource and environmental accounts described above support the implementation of these models.

Building complex policy models may be an expensive luxury in many developing countries, however. For these countries, rapid assessments of resource depletion and the value of environmental degradation, placed in the savings and wealth framework presented above, will guide policy makers aiming for sustainable development. Green national accounting aggregates, including Hicksian income (EDP) and genuine savings, place natural resources and the environment in an economic context that is otherwise lacking in developing countries.

Growth theory provides the intellectual underpinning for expanded national accounting and, through the measure of genuine saving, an indicator of when economies are on an unsustainable development path. This theory points in useful directions for countries concerned with sustainable development.

Far from being a mere theoretical possibility, there is abundant evidence for countries whose policy mix results in negative genuine saving rates. While the World Bank estimates for 1997 are emphasised here, previous studies such as Hamilton and Clemens (1999) and Atkinson et al. (1997) have shown this to be true over decades as well.

The evidence suggests that, while resource-dependent economies are potentially sustainable if resource rents are invested in other productive assets (including human capital), many of these economies have not chosen this path. The results presented here show distinctive patterns of genuine savings across regions and country income groups.

However, as the example of Southeast Asia in 1997/98 shows, robust genuine savings do not necessarily lead to a smooth development path. Some of the broader lessons from the financial crisis in Southeast Asia concern the rates of return that were achieved with these savings; many savings investments were yielding zero or exceedingly small returns. So the lessons to be drawn from the analysis of genuine saving must go beyond the level of saving, to a concern with the quality of the investments that are made with these savings.

The genuine savings analysis raises an important set of policy questions that goes beyond the traditional concern with the macro and macroeconomic determinants of savings effort. The questions of rent capture, public investments of resource revenues, resource tenure policies, and the social costs of pollution emissions, are equally germane in determining the overall level of saving, although, it is clear that monetary and fiscal policy remain the big levers.

This analysis also provides a practical way for natural resource and environmental issues to be discussed in the language that Ministries of Finance understand. This may prove to be an important advantage as many resource-dependent economies struggle to achieve their development goals.

End Notes

1. Pearce and Atkinson (1993) presented the first cross-country empirical work on adjusted savings measures.
2. Hartwick (1990) and Mäler (1991) extended the growth-theoretic approach to national accounting to deal with environment and natural resource issues, while Hamilton (1996) looked in detail at accounting for pollution.
3. Human capital provides a type of endogenous technical progress. Cf. Weitzman and Löfgren (1997), who deal with exogenous technical change.
4. More elaborate models, such as that of Vincent (forthcoming), deal explicitly with the rotation age of forest resources, so that for a given stand there is several years of net appreciation of the stock, followed by depreciation at the point of harvest.
5. This if often, confusingly, termed the 'net price' method, while the simple present value is, equally confusingly, termed the 'user cost' approach. Both methods can be shown to measure user costs of resource extraction under appropriate assumptions.
6. This should be termed more accurately the willingness to pay to reduce the risk of death. Most governments have an 'official' value of a statistical life that is used in establishing safety standards for public infrastructure such as highways.
7. While strong sustainability was originally formulated with critical stocks such as rainforest or the ozone layer in mind, the analytical issues associated with a critical stock of pollution are precisely the same.
8. Atkinson *et al.* (1997) derive the more general expression when total wealth is a function of population.
9. It must be recalled, however, that the total rent approach to measuring depletion tends to exaggerate the value of depletion, particularly for countries with very large resource endowments.
10. The question is also germane for unsustainable forest harvest programs.

References

Alfsen, K.H. (1993), *Natural Resource Accounting and Analysis in Norway,* Conference on Medium-Term Economic Assessment, Oslo, 2–4 June.

Aronsson, T., P.O. Johansson and K.G. Löfgren (1994), 'Welfare measurement and the health environment', *Annals of Operations Research,* **54,** 203–15.

Atkinson, G., R. Dubourg, K. Hamilton, M. Munasinghe, D.W. Pearce and C. Young (1997), *Measuring Sustainable Development: Macroeconomics and the Environment*, Cheltenham: Edward Elgar.

Ayres, R.U. and A. Kneese (1969), 'Production, consumption and externality', *American Economic Review,* **59,** 282-97.

Dasgupta, P., and K.-G. Mäler (forthcoming), 'Decentralization schemes, cost benefit analysis, and Net National Product as a measure of social well-being', *Environment and Development Economics.*

de Haan, M., S.J. Keuning and P. Bosch (1993), *Integrating Indicators in a National Accounting Matrix Including Environmental Accounts* (NAMEA), NA 060, 1993, Voorburg: Central Bureau of Statistics.

El Serafy, S. (1989), 'The proper calculation of income from depletable natural resources', in Y.J. Ahmad, S. El Serafy and E. Lutz (eds), *Environmental Accounting for Sustainable Development,* Washington: The World Bank.

Fankhauser, S. (1995), *Valuing Climate Change: The Economics of the Greenhouse,* London: Earthscan.

Gelb, A.H. (1988), *Oil Windfalls: Blessing or Curse?* New York: Oxford University Press.

Hamilton, K. (1994), 'Green adjustments to GDI', *Resources Policy,* **20** (3), 155–68.

Hamilton, K. (1996), 'Pollution and pollution abatement in the national accounts', *Review of Income and Wealth,* **42** (1), 13–33.

Hamilton, K., G. Atkinson and D.W. Pearce (1998), 'Savings rules and sustainability: selected extensions'. Presented to the World Congress oil Environmental and Resource Economics, Venice, June.

Hamilton, K., and M. Clemens (1999), 'Genuine savings rates in developing countries, *World Bank Economic Review,* **13** (2), 333–56.

Hartwick, J.M. (1990), 'Natural resources, national accounting and economic depreciation', *Journal of Public Economics,* **43**, 291–304.

Hicks, J.R. (1946), *Value and Capital,* 2nd Ed., Oxford: Oxford University Press.

Kunte, A., K. Hamilton, J. Dixon and M. Clemens (1998), 'Estimating national wealth: methodology and results', Environment Department Papers, Environmental Economics Series No. 57, Washington: The World Bank.

Leontief, W. (1970), 'Environmental repercussions and the economic structure: an input–output approach', *Review of Economics and Statistics,* **52**, 262–71.

Mäler, K.-G. (1991), 'National accounts and environmental resources', *Environmental and Resource Economics,* **1**, 1–15.

Pearce, D.W., and G. Atkinson (1993), 'Capital theory and the measurement of sustainable development: an indicator of weak sustainability', *Ecological Economics,* **8**, 103–8.

Pearce, D.W., A. Markandya and E. Barbier (1989), *Blueprint for a Green Economy,* London: Earthscan Publications.

Pearce, D.W., K. Hamilton and G. Atkinson (1996), 'Measuring sustainable development: progress on indicators', *Environment and Development Economics,* **1**, 85–101.

Pezzey, J. (1989), 'Economic analysis of sustainable growth and sustainable development', Environment Dept. Working Paper No. 15, Washington: The World Bank.

Repetto, R., W. Magrath, M. Wells, C. Beer and F. Rossini (1989), *Wasting Assets: Natural Resources in the National Accounts,* Washington, World

Resources Institute.
United Nations (1993a), *System of National Accounts 1993,* New York: United Nations.
United Nations (1993b), *Integrated Environmental and Economic Accounting,* Series F No. 61, New York: United Nations.
Victor, P.A. (1972), *Pollution, Economy and Environment,* London: Allen and Unwin.
Vincent, J., T. Panayotou and J.M. Hartwick (1997), 'Resource depletion and sustainability in small open economies', *Journal of Environmental Economics and Management,* **33**, 274–86.
Vincent, J. (forthcoming), 'Net accumulation of timber resources', *Review of Income and Wealth.*
Weitzman, M.L. (1 976), 'On the welfare significance of national product in a dynamic economy', *Quarterly Journal of Economics,* **90**, 156–62.
Weitzman, M.L. and K.-G. Löfgren (1997), 'On the welfare significance of green accounting as taught by parable', *Journal of Environmental Economics and Management,* **32**, 139–53.
World Bank (1997), *Expanding the Measure of Wealth: Indicators of Environmentally Sustainable Development,* Environmentally Sustainable Development Studies and Monographs Series No. 17, Washington: The World Bank.
World Bank (1999), *World Development Indicators 1999.* Washington: The World Bank.

8

The Standard Welfare Economics of Policies Affecting Trade and the Environment

Kym Anderson

The greening of world politics in recent years has brought forward, among other things, claims and counter-claims as to the effects of trade and hence trade policy on the environment. Environmentalists have also propagated suggestions for government intervention to slow the degradation of the natural environment which, if adopted by one set of countries, would affect the trade, environment and welfare of other countries as well. There is thus a growing demand for a better understanding of the welfare economics of the various interactions between trade and the environment, and between environmental and trade policies.

This chapter draws on the simplest economic theory that is able to demonstrate the linkages between trade, the natural environment and welfare. Using partial-equilibrium, comparative-static analysis, it seeks to show how the environmental and trade policies of one country (or set of countries) can affect its own and the rest of the world's environment, trade and social welfare.

After listing the basic assumptions to be used (some of which are relaxed during the course of the analysis), the following key questions are addressed:

- What are the effects on the environment and social welfare of a country opening up to or liberalizing its trade?
- How are these effects altered when the country also adopts appropriate environmental policies?
- To what extent are trade policy instruments able to substitute

for first-best means of overcoming environmental externalities?

- How do changes in environmental and trade policies abroad affect this open economy's environment and welfare?

Initially the analysis focuses on an economy small enough for its production and consumption activities not to have significant effects on the rest of the world. The large economy (which could include a group of several similar small economies) is then analysed, to show the effects of its own and its trading partners' policies not only at home but also abroad. The final section of the chapter summarizes the results and mentions some qualifications that need to be kept in mind.

8.1 BASIC ASSUMPTIONS

Consider an economy in which one group's production (or consumption) of a good begins to impose an externality on others through its effect on the natural environment. That is, this economic activity involves a marginal divergence between the private and social cost of production (or benefit from consumption), by adding to other producers' costs or reducing the aesthetic pleasure provided by nature. Such a divergence may have arisen because of new knowledge about the activity's pollutive effect becoming available or because preferences for a cleaner environment have strengthened, or because a threshold level of pollution has been reached which triggers concern for the environment. Whatever the reason, assume that property rights are not defined clearly enough and/or that there are high transactions costs which prevent the full internalization of the externality.[1] For simplicity, assume also that there are no administrative or by-product distortionary costs of collecting taxes or disbursing subsidies and that the income distributional effects of such transfer policies can be neglected. Assume too that producers, consumers and policy-makers are well informed and can appropriately value the aesthetic or material cost (or benefit) of the externality involved.[2]

The externality is assumed to result from the production (or consumption) activity itself, not from the use of a particular process. This means that a tax-cum-subsidy on production (or consumption) is equivalent to a tax-cum-subsidy on the source of the externality and thus is the optimal environmental policy instrument for correcting this divergence.[3] Initially assume also that the externality results from the production (or consumption) of just one good, and that there are no distortionary policies affecting other markets in this economy—although both these assumptions are relaxed later in the analysis.

To simplify the exposition a negative externality will be referred to as pollution, but it should be kept in mind that the analysis is equally relevant for all types of externalities. We begin with the small-country case, in which international prices are given and the effects on the rest of the world can be ignored, before turning to the large-country case. As is appropriate in this comparative-static analysis, changes in tastes and technological changes are not considered, nor is international factor mobility.

8.2   THE SMALL-COUNTRY CASE

The marginal private benefits from consuming and the marginal private cost of producing a particular good are represented by the D and S curves, respectively, in Figure 8.1. Production (but not consumption) of this good is pollutive; the marginal social costs of production is represented by S'. (In practice, social and private costs may not diverge until the level of production has passed a certain threshold, and may not diverge increasingly over all output ranges, but the exposition is simpler and loses no generality by depicting all curves as linear with the divergence beginning with the first unit of production.) These curves fully incorporate the feed-back effects of changes in production and/or consumption of this good on the markets for other goods, for productive factors and for foreign currencies in this economy. The price axis refers to the price of this good relative to all other prices in the economy (which are assumed to remain constant throughout).

Given the above assumptions, OQ would be the equilibrium level of production and consumption in the absence of both a pollution tax and international trade, yielding net social welfare equal to the difference between areas *abe and ade*.[4] Suppose the economy's trade policy were to change from autarchy to free trade. Then if OP_o was the international price at this small economy's border, as in Figure 8.1(a), production would fall to OQ_m consumption would rise to OC_m and Q_mC_m units would be imported. Net social welfare would now be *abfg-ahg,* so the welfare gain from opening up to trade is *defgh*. Not only is this gain from trade positive but it is greater than it would have been in the absence of a negative production externality, by shaded area *degh* (the difference between the private and social cost of producing Q_mQ units domestically). On the other hand, if OP_1 was the international price, as in Figure 8.1(b), the country would become an exporter of C_xQ_x units of this good under free trade. In that case net social welfare would be *abik-amk,* so the welfare effect of opening up to trade is *eik-dekm,* which may be positive or negative. The ambiguity arises because the gain from trade in the absence of an externality associated with the extra production (*eik*) is more or less than offset by the uncharged cost of polluting the environment by producing an additional QQ_x units of this product (*dekm*).

In short, *liberalizing trade in this good whose production is pollutive improves the small country's environment and welfare if, following the policy change, the country imports this good; but should this good be exported, the environment is worsened and so welfare in this small country may or may not increase in the absence of a pollution tax* (Proposition 1).

What difference would it have made if this country had in place its optimal environmental policy both before and after trade was liberalized? In this case the optimal policy would be a production tax equal to the vertical distance between S and S' at the output level at which the marginal social cost of production equals the marginal social benefit. In the autarchy case the optimal intervention would be a tax of *cn* per unit produced, which would induce output of OQ_o rather than OQ in Figure 8.1 (a) (yielding a welfare benefit represented by the dark shaded area *cde,* the difference between the social cost and benefit of those Q_oQ units).[5] In the free trade case the optimal production tax would be *qr* if OP_o was the international price (Figure 8.1 (a)), or *js* if OP_1 prevailed (Figure 8.1(b)), both of which would curtail production somewhat and thereby improve the environment and raise welfare relative to the situation where no pollution tax existed. In the exporting case, for example, production would decline from OQ_x to OQ_x' in Figure 8.1(b), improving social welfare by shaded area *jkm*. The gain from opening up to trade in the presence of such optimal environmental policies is represented by *qcf* in Figure 8.19(a) in the case where this good would be imported or *cij* if it were to become an exportable. If the optimal pollution policy was not in place before trade liberalization, the gain from reform would be even greater, by *cde*. That is, *even in the export case there is an unambiguous welfare gain from trade for a small country provided something approaching the optimal environmental policy also is introduced, despite the fact that the environment is more polluted when production expands to supply exports* (Proposition 2).

The above analysis of the move from autarchy to free trade is easily modified to examine the effects of reducing or removing less-than-prohibitive trade taxes. In that case the initial domestic price is closer to the free-trade price and domestic production does not equal domestic consumption, but the conclusion is the same: lowering barriers to imports of this good will improve the environment and welfare in

(a) Importable

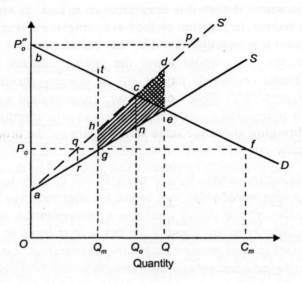

(b) Exportable

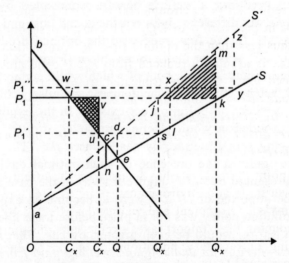

Fig. 8.1: Effects of opening up a small economy to trade
in a product whose production is pollutive

this small country especially with, but even without, the optimal pollution tax, whereas if the good is an exportable, then lowering export barriers will worsen the environment but will nonetheless improve welfare so long as something close to the optimal environmental policy is in place.[6]

Suppose there are prohibitive administrative costs involved in imposing the optimal environmental policy instrument, namely, a production tax. In the case where the good is exportable an export tax could be used instead to reduce degradation of the environment, but it would be less efficient than a production tax. This can be seen from Figure 8.1(b). An export tax *of js* would lower the price producers receive and hence domestic consumers pay by $P_1{'} P_1$, thereby shrinking exports from $C_x Q_x$ to $C_x{'} Q_x{'}$. This would ensure the marginal social cost of production is lowered to its marginal benefit (the international price OP_1), so bringing about the same welfare gain on the production side as an equally large production tax, namely shaded area *jkm*. But the export tax also imposes a by-product cost of distorting the consumption side of this market. By setting the price to consumers below the opportunity cost of OP_1, the deadweight welfare cost of encouraging an extra $C_x C_x{'}$ units of consumption is shaded area *iuv*. This inferior environmental policy instrument therefore reduces environmental degradation as much as a production tax of the same rate, but at higher cost than the optimal pollution tax instrument. Indeed if *iuv* exceeds *jkm* this second-best policy instrument would also be worse than no intervention, despite the reduction in environmental degradation brought about by the cut-back in production to $OQ_x{'}$. The optimal second-best export tax rate is thus less than the optimal first-best production tax rate: as the export tax is gradually reduced from $P_1{'} P_1$, the area of triangle *iuv* is reduced more than the truncated triangular area *jmk*.[7] More generally, *trade taxes-cum-subsidies can be used to reduce environmental degradation by a given amount, but they will improve welfare less than a more direct tax on the source of pollution and may even worsen welfare* (Proposition 3).

To this point the analysis has focused on the effect of this small country's own trade and environmental policies. But how would this open economy's environment and welfare be affected by other large countries reducing their import taxes/export subsidies on this product or introducing or raising a pollution tax on its production? These policy changes abroad—like an exogenous expansion in overseas excess demand—would raise the international price of this product. Would this increase or decrease the benefit our small economy enjoys from being open to free trade?[8] If the international price rose from OP_1 to $OP_1{'}$ in Figure 8.1(b), the welfare benefits from being an exporter would rise by *ijxw* if the optimal pollution tax was in place and raised accordingly from *js* to *xl*. If no pollution tax existed, welfare would change by *ikyw-mkyz*, which might be positive or negative. That is, *the welfare benefit to a small country of adopting a free trade policy would be enhanced if the international price of its exportable rose—*

notwithstanding the greater environmental damage caused by greater production—so long as it had a sufficiently small divergence between its marginal social and private cost curves or applied a pollution tax close enough to the optimal rate (Proposition 4). The pollution tax needs to ensure that the new vertical area between the S and S' curves and bounded by the quantities produced before and after the change does not exceed the new horizontal area between the S and D curves and bounded by OP_1 and OP_1''.

On the other hand, if this small country is and remains an importer of this product at the new international price, its welfare benefits from being open to free trade rather than autarchic (*defgh* in Figure 8.1(a) if there is no production tax) would be truncated as the price line *fg* moves up and the vertical line *gh* moves to the right. The benefit if the optimal production tax was in place and adjusted accordingly (*qcf* in Figure 8.1(a)) would also be truncated as the base of that triangle, *qf*, moved up with the rise in the international price.

There is the possibility, however, that the international price rise causes this country to switch from being an importer to an exporter of this product. Should that change be sufficiently large, the country's welfare gain from being open to trade rather than autarchic could turn out to be *enhanced.* This would happen if, for example, the international price rose by P_oP_o'' in Figure 8.1(a) and the optimal production tax was in place and adjusted accordingly, and *bcp exceeded qcf.*

The practical significance of this result is important: *even a country importing a product whose international price rises because of the imposition of environmental policies or a reduction in protectionism abroad need not be made worse off—even though its own environment will be harmed by the increased output and hence pollution stimulated by that price rise—if that country were to switch sufficiently from being an importer to an exporter of this product and had a sufficiently small divergence between its marginal social and private cost curves or applied a pollution tax close enough to the optimal rate* (Proposition 5).

Indeed the possibility of an importing country gaining from an international price rise need not even require that country to switch to exporter status if one or two of the basic assumptions listed in Section 8.1 above does not hold. One assumption is that other markets in this economy are not distorted by policies; another is that there are no externalities elsewhere in the economy. Consider relaxing each of these assumptions in turn.

If the sector under discussion is implicitly favoured or discriminated against by negative or positive government assistance to other sectors, then the international price line in Figure 8.1 has to be redefined. Recall that the price axis refers to the price of this sector's product relative to the prices of all other products (which are assumed to be constant). If the composite price of other products is, say, higher domestically than internationally because of protection in this economy's other import-competing sectors, then the relative price of this sector's product in the domestic market is lower than it is in the international market.

To see the significance of this, consider Figure 8.2 in which *ED and ED'* are the economy's private and social excess demand curves for this product (the horizontal difference between the supply and demand curves in Figure 8.2), and OP_o is the (relative) price of purchasing a unit of this product domestically, given the protection afforded to other sectors. The international relative price of this product, however, is necessarily higher than the domestic relative price in this distorted economy. If it is above *e*, say *OP*, the economy would be an exporter of this product under free trade, even with the optimal pollution tax in place. In this setting it turns out that should the international price of this sector's product rise, this economy with its optimal pollution policy in place will gain even though it may remain a net importer of this product.

The reasoning is as follows. Suppose the international price increases from *OP* to *OP* and this increase is fully transmitted so that the domestic price rises from OP_o to OP_o'. The protectionist policy reduced welfare in this economy (with its optimal pollution tax) by *hcf* when *OP* and OP_o prevailed. Now with *OP'* and OP_O' prevailing, the welfare loss from protection is *nkj*. But if the social excess demand curve is linear and the international price change has been fully transmitted, areas *hcf* and *nkj* are identical. However, the rise in the international price itself adds to the country's welfare, because of the country's comparative advantage at price *OP* and hence *OP'*. If the economy was undistorted, it would have benefited by *adhn* from the price rise, which is the producer gain net of the consumer loss. In the presence of the distortionary protectionist policy, the producer gain net of the consumer loss is a negative value, *fgrj*, but off-setting that is the reduction in the implicit import subsidy to foreigners which is a saving of *bqjk* (= *cfts*). The sum of these turns out to be equivalent to *adhn*.[9] That is, even though the country remains a net importer of this product, and even though its environment would be more degraded as domestic output of this good expands, the rise in its international price would (if P is above e in Figure 8.2) boost welfare in this distorted economy.

The second assumption that, when relaxed, can lead to a conclusion that an importing country could gain from an international price rise has to do with externalities in the markets for other products in this economy. If other sectors also pollute, then diverting resources from this sector to others is less socially useful than is suggested in Figures 8.1 and 8.2. That is, the marginal social cost curve S' (or the social excess demand curve ED') is closer than depicted to, or may even be below, the marginal private cost curve S (or the private excess demand curve ED). If in Figure 8.2 ED'' was the true social excess demand curve, on the assumption that other industries pollute more than this one, then with optimal pollution taxes in place this good would be exported rather than imported if the international relative price was OP_o. Even without pollution taxes the country would gain (by gmur) rather than lose from an increase in the international price from OP_o to OP_o' that was fully transmitted domestically, again despite the fact that output and hence pollution from this industry would increase. In this case, however, the environment would improve because pollution from other industries would decrease by more than was added by this industry. (Had the optimal pollution tax been in place the country

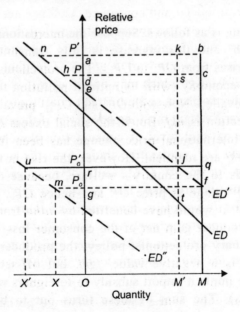

Fig. 8.2: Effects of an international price rise for a small distorted economy

would have been an exporter of this product, and its exports would have increased from OX to OX' when the international price rose from OP_o to OP_o' and the pollution tax was adjusted to its new optimal rate. The welfare gain from the international price rise in this case would be *gmur*, the same as when there are no pollution taxes.)

In short, these two illustrations using Figure 8.2 demonstrate that *even an economy which remains a net importer of a product whose international price has risen could benefit from that price rise if (a) domestic production of that product was being discouraged sufficiently by assistance to other sectors of the economy or (b) other sectors were more pollutive than this sector so that attracting resources into this sector improved the country's environment sufficiently* (Proposition 6).

Before moving on to consider the large-country case, it should be noted that negative consumption externalities, such as from burning fossil fuels, can be analysed in a similar fashion to production externalities. If a country had no pollution taxes and were to open up to or liberalize trade in such products, its environment and welfare would improve if it became an exporter as the higher domestic price would curtail domestic consumption. If it became an importer, on the other hand, consumers would gain more than producers lose but pollution would increase so welfare might increase or decrease in the absence of a pollution tax. This can be seen from Figure 8.3. In autarchy Q units of the domestic product would be produced and consumed, yielding net social welfare represented by *abj-abg* in the absence of a pollution tax. Under free trade, the country would import $Q'C'$ units at the international price OP, yielding net social welfare of *acej-ack*. The gain from trade in this case is thus *deh-bckg*, which may be negative. Should the optimal pollution policy be in place and adjusted accordingly when trade is liberalized, however, welfare must improve even if the country becomes an importer. The optimal level of consumption then would be OQ_o under autarchy and OC_o under free trade, with the net welfare gain from opening up to import $Q'C_o$ units being represented by the unambiguously positive area *deh*. That is, *liberalizing trade in a good whose consumption is pollutive improves the country's environment and welfare if the country exports that good, but would worsen the environment and therefore may reduce welfare if the good is imported unless a pollution tax close enough to the optimal rate is in place* (Proposition 7).

What if consumption of the foreign imported product was not as polluting as the domestic product? An example might be coal, where local coal contains more sulphur and so adds more acid rain than does the burning or foreign coal (as may be the case in Europe—see Newbery, 1990). Suppose in the extreme that the rest of the world is unworried by acid rain and so is indifferent as to the sulphur content of the coal it consumes, but for this country there is a social preference for imported coal which, for the sake of exposition, is assumed to be sulphur-free and to involve no other externalities. In this case it is possible that even in the absence of an appropriate environmental policy, import liberalization could improve both the environment and welfare because under free trade the country would import all its needs from abroad at price OP in Figure 8.3 and export its own product abroad, also at price OP (given the assumption that foreigners are indifferent as to the sulphur content of the coal they burn). The consumer surplus with free trade is acf and producer surplus is jfe, yielding net social welfare of $acej$. There is a clear gain in both environmental quality and net social welfare from opening up to trade in this case, the welfare gain being equal to $aceh$ plus hgb. While this gain would be less if imported coal contributed some rather than no sulphur pollution, or if a sulphur emission tax had been in place before trade was opened up, this example serves to make the point that import liberalization of goods whose consumption is pollutive need not necessarily add to pollution. Or, to put it another way, where consumption of the imported product is less pollutive than consumption of the domestic substitute, an import barrier may be more rather than less costly to this country than is conventionally measured without accounting for pollution (Proposition 8).

8.3 THE LARGE-COUNTRY CASE

How does the above analysis change when it is no longer assumed that the country's production and consumption do not affect the markets and environment of other countries? One change is that the export demand and import supply curves are no longer horizontal lines at a given international price as drawn in Figures 8.1 to 8.3, since this country's activities are sufficiently large to affect the price in the international market. Another fundamental change is that the divergence between the country's social and private excess demand curves alters, because its actions affect pollutive activities abroad which may spill over into this country's environment. Together these changes ensure that the effect of the country opening up to or liberalizing trade are somewhat different in the large-country case than in the small-country case. They also ensure that policy changes in one large country (or group of small countries) may cause other countries to change their

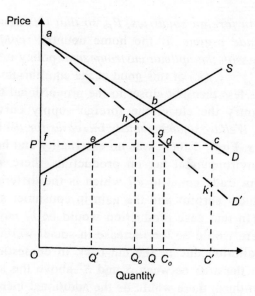

Fig. 8.3: Effects of opening up a small economy to trade in a product whose consumption is pollutive.

policies too, so a wide range of outcomes is possible as the following examples illustrate.

Consider first a product whose production and consumption involve no international environmental spillovers. In the home country production of this good is more pollutive than would be production of the next most profitable good. Suppose in the rest of the world (call it the foreign country for simplicity), however, that this product is no more pollutive than other products. Hence the social and private marginal cost curves coincide abroad but diverge at home. An example might be agriculture, where in the advanced industrial (home) economy the alternative use of mobile farm resources is in the relatively unpolluting service sector whereas in the less-developed economies their alternative use would be in equally polluting smoke-stack manufacturing industries. Moreover, the advanced economy may use more chemical inputs and more intensive livestock methods than the poorer, more agrarian economies, and be relatively densely populated, in which case at the margin its farmers would be more pollutive than are farmers in less-developed economies.[10]

The effects of the large home country opening up its trade in this setting can be seen from Figure 8.4, where *it is assumed that costs of transporting this product between the home and foreign countries are zero and that under autarchy, the price at home, P_H, would be* *above the price in foreign countries, P_F, so that this item is imported under a free trade regime.* If the home country *replaced its self-sufficiency policy with the optimal environmental* policy it would import OT units (= $Q_m C_m = C_x Q_x$) of this good at the equilibrium international price OP. This is less than but closer to the proportional trade increase for a small country the closer the foreign supply curve, FS, is to being horizontal. *Welfare unambiguously increases for the home country for three reasons.* First, if there was no externality and hence no need to impose an environmental tax on production, there would be the normal gain from trade equal to *efj* which is the difference between the loss in producer surplus and the gain in consumer surplus in the home country. (In that case production would be *hj* more units than Q_m) Second, there would be the increase in social welfare associated with reduced pollution due to the cut-back in domestic production. This is equal to the area between S and S' above the segment *je* of the S curve. And third, there would be the additional increase in social welfare (net of the loss of producer surplus *hgj*), associated with reducing production even further, to Q_m, by imposing the optimal environmental tax on production. That is the area between S and S' above the line *hj*. The total gain, shaded area *defh,* is smaller than it would have been without the international price rise (the small country situation) by *fkmh*, but it is certainly positive. Moreover, there is a welfare gain of shaded area *rst* and (by assumption) no extra environmental degradation in the foreign country following this trade liberalization. It follows that even in a situation where the foreign country's expanded production added to local and global pollution, both countries could still be better off as a result of the home country's reduction in import protection.

Now consider a case of liberalizing export trade, and explicitly include international environmental spillovers. Figure 8.5 depicts markets for a good (steel, for example) whose production involves the same by-product emissions in the two countries. Emissions in each country are partly transmitted to the other and assume these emissions are considered pollutive by the home country but not by the poorer foreign country. Given the foreign country's excess demand curve, the home country is assumed to be able to assess how much production tax at home not only would reduce home emissions but also would increase pollution emanating from abroad of concern to home-country citizens (either because it is blown or flows to the home country or because, in their opinion, it affects the global commons adversely). The fact that taxing production at home encourages production abroad once this economy is opened ensures that the marginal social cost of domestic

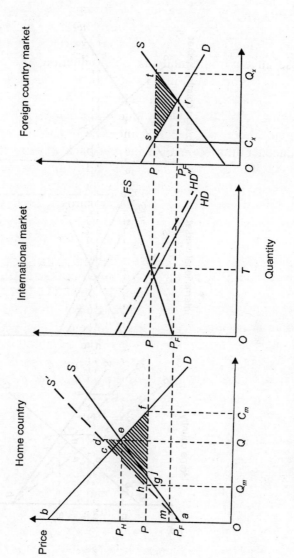

Fig. 8.4: Effects of opening up a large economy to imports of a product whose production is pollutive

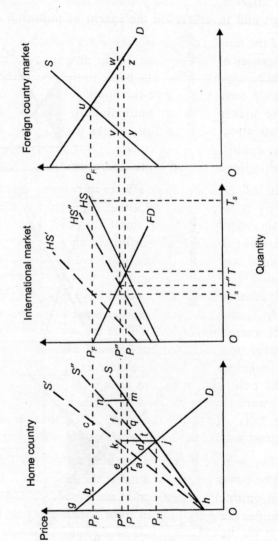

Fig. 8.5: Effects of opening up a large economy to imports of a product whose production is pollutive

production drops somewhat below what it would be if the economy was small, say to S''. The vertical gap between S' and S'' would be larger, the greater the extent to which the benefits to home-country citizens of decreased emissions from domestic production are perceived to be offset by increased pollution from abroad, which in turn depends in part and inversely on the extent of pollution taxes abroad.

If the home country in Figure 8.5 opens up to trade, it will become an exporter of this product. In the absence of a pollution tax OT units would be exported from the home to the foreign country at international price OP (which is above the home country's autarchic price of OP_H). In the presence of an optimal pollution tax in the home country, liberalization would result in less trade (OT'') and the international price would be higher at OP''. But in both cases the new price at home is lower than the pre-liberalization foreign price of P_F.

In this case the trade effects of opening up to free trade when the country is large could be either proportionally larger or smaller than for an otherwise-similar small country, depending on the offsetting effects of (a) a less-than-infinitely elastic export demand curve facing the home country on the one hand and (b) the propensity for transfrontier pollution on the other hand. In the case illustrated in Figure 8.5 the small country facing OP_F would have exported OT_s units in the absence of a pollution tax, whereas the large country only exports OT at the lower international price OP. If the optimal pollution tax was applied, however, the small country would have exported OT_s' units at price OP_F, whereas the larger country exports OT'' at price OP''. That is, in the case illustrated the change in the effect on trade of facing a downward-sloping rattler than horizontal export demand curve is more than fully offset by the trade effect of the drop in the domestic marginal social cost curve because of transfrontier pollution from abroad.

The welfare effect of opening up to free trade also could in this case be either greater or smaller the larger the country. In the presence of an optimal environmental policy before and after opening up, for example, the welfare gain from opening to free trade is *abc* in Figure 8.5 if the home country is small (and therefore has no effect on the foreign country's pollution). When it is large, however, the welfare gain is *def,* which may be more or less than *abc*. Nor is it clear whether under free trade the environment will be cleaner for the larger country than for the small country. This is because output may expand more or less in the large country than the small country, but even if it expands less and so means a smaller increase in domestic emissions for the large country, there may be extra spillover of foreign pollution in the latter case which could be enough to offset that difference.

As in the small-country case, trade liberalization for this export commodity (as distinct from an import-competing product) need not necessarily increase welfare in the large country if there is no pollution tax in place. Before trade is allowed and when no pollution tax applies, social welfare from producing and consuming this product is represented Figure 8.5 by *dgh-djt*. After trade is opened, social welfare in the absence of a pollution tax would be *qrgh-qmn*. Thus the change in the home country's welfare from opening up fully to trade in the absence of a pollution tax is *qrjt-qmn,* which could be positive or negative. Should the optimal pollution tax be introduced when trade is opened, however, welfare certainly improves at home (by *eftj*).[11] The superior outcome in the latter case arises not only because marginal social costs and benefits are brought into line by the pollution tax (contributing *qmm*) but also because that tax improves this large country's terms of trade (adding *refq*).

Given its lack of concern for the environment, the foreign country in Figure 8.5 gains from this trade; it would gain by *uyz* if no pollution tax is applied in the home country or by the lesser area *uvw* if the optimal pollution tax was introduced in the home country when its trade was opened up. That is, that pollution tax itself reduces welfare in the foreign country in this case where that foreign country is and remains an importer of this product.

These results expose the intriguing contradictory possibility that a large country (or group of small countries) that exports a polluting product but is unconcerned by that pollution might nonetheless benefit from taxing the production of that good and might at the same time improve welfare for its trading partners who care more for the global environment. This could happen when the gain in welfare of the 'greener' country group due to the cleaner global environment more than offsets the global loss in welfare due to reduced consumption of goods, and the net gain in global welfare (which is the difference between these two effects) is shared between the exporting and the importing country groups. Examples that come to mind are taxing the carbon content of fossil fuels at the point of production, or taxing logging activities to compensate for the foregone carbon absorptive capacity of depleted forests.

To illustrate this possibility, consider an extreme case in which exporting countries do not consume and importing countries do not produce the product in question. Then the international market for this good, depicted in Figure 8.6, is also the global market. The curve *MD* is the demand in the importing countries and *XS* is the supply in the exporting countries, while the vertical distance between *XS and XS'* reflects the negative aesthetic value importers place on an extra unit of production. If the importers were successful in persuading the exporters to place a tax of *bd* per unit on production of this good, the price would rise from *OP* to *OP'*. The importers' loss in consumer welfare is *bcgh,* but their welfare gain from reduced environmental degradation is *bdcj* which conceivably could exceed *bcgh* and thereby make them better off. At the same time exporters are better off as well (even assuming they get no aesthetic pleasure from a cleaner environment) so long as the reduction in producer surplus, *cdfg,* is less than the production tax revenue, *bdfh.* In that case the global gain, *bcj,* would conveniently be shared between the two groups of countries without explicit (non-market) foreign transfers.

Clearly administrative problems exist in collecting producer/exporter taxes and distributing those tax revenues between producing countries. The strategy adopted by OPEC of course has been to try to agree on national production quotas, but even then there are policing costs because of the free-rider incentive to produce above quota. Moreover, as Chapter 4 in this volume by Snape points out, importing countries

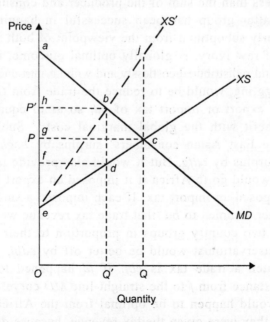

Fig. 8.6: Effects of taxing production or consumption
to improve the global environment

alternatively or additionally could tax consumption and thereby possibly improve their own and global welfare, but at the expense of the exporters in this situation.

So far, only a two-country world has been considered. International cooperation becomes more complex when three or more countries/ country groups are involved, as the controversy over trade in raw ivory exposed (see Barbier *et al.*, 1990). Figure 8.6 can be used to illustrate crudely the main features of that controversy. Africans supply raw ivory almost exclusively for export. East Asians are the main demanders of that ivory, and of course they have no domestic supply source. Hence *XS and MD* in Figure 8.6 represent both the domestic and the excess supply and demand curves for these two groups, respectively. But there is a third group of people, mainly in Western Europe and North America, who believe this trade is causing the African elephant herd size to be smaller than they feel is desirable. The distance between *XS and XS'* in Figure 8.6 represents the marginal 'cost' to that third group of the slaughter required to supply the raw ivory trade (their willingness to pay).[12]

In this situation free trade in ivory would provide African suppliers with a welfare gain of *ceg* and East Asian consumers with a welfare gain of *cag*. But the conservationist group would lose *cej*, which may be more or less than the sum of the producer and consumer surpluses (*cea*). The latter group has been successful in banning this trade, which is clearly suboptimal from the viewpoint of both producers and consumers of raw ivory. A globally optimal outcome, if taxes could be collected and redistributed costlessly and without introducing distortions such as smuggling, would be to reduce the trade from *OQ to OQ'* by imposing an export or import tax of *bd,* so as to equate the global marginal benefit with the global marginal cost.[13] Such a trade tax would reduce East Asian consumers' surplus by *bcgh* and African producers' surplus by *cdfg*. But it would also provide tax revenue of *bdjh*, which would go to Africa if it imposed an export tax or to East Asia if it imposed an import tax. If each imposed a smaller trade tax which together summed to *bd,* that trade tax revenue would be shared between the two country groups in proportion to their tax rates. As well, the conservationist would be better off by *cdbj,* so the global gain from such a trade tax is *cbj*. If *fd* happened to be half the horizontal distance from *f* to the straight-line *MD* curve, an export tax of *bd* also would happen to be optimal from the African producers' viewpoint if they were given the tax revenue, because *d* would be the point of intersection of the marginal revenue curve from point *a*, given the average revenue curve *MD*. Or if *hb* was half the horizontal distance from *h* to the *XS* curve, an import tax of *bd* would happen to be optimal from the East Asian consumers' viewpoint if they were given the tax revenue. Clearly in this case the conservationists could seek support to restrict trade to *OQ'* from either producers or consumers, but the excluded group would oppose or retaliate in response to that restriction. And both producers and consumers of ivory would oppose restricting the trade to less than *OQ'*. Conservationists have been prepared to insist on a ban on raw ivory trade in large part because they have not been required to compensate the losers. If they were, they would need to be able to pay the equivalent of *cea,* and even if the benefit to them of the ban (*cej*) exceeded that loss to ivory suppliers and consumers, it might not exceed it sufficiently to overcome free-rider and other administrative costs of collecting that revenue from supporters of this conservationist policy and redistributing it to the relevant producers and consumers.

8.4   SUMMARY, QUALIFICATIONS AND IMPLICATIONS

Several key points can be summarized from the above analysis concerning the effects of trade and environmental policies on the environment, trade and welfare. First, opening up trade in a good whose production is relatively pollutive improves the environment and welfare of a small country if, after opening up, this country imports the good in question. Should this good be exported, however, trade expansion through trade liberalization worsens a small country's environment, and so welfare may or may not increase in the absence of a pollution tax. And conversely if the pollution is generated from consumption. Even the small country exporting (importing) this good whose production (consumption) is relatively pollutive would gain unambiguously from trade liberalization if it simultaneously ensured that something close to the optimal environmental policy was in place, despite the fact that its environment may be more polluted under a liberal trade regime.

Second, trade taxes-cum-subsidies could be used to reduce environmental degradation by a given amount, but they would improve a country's welfare less than would a more direct tax on the source of pollution—indeed they may even worsen welfare.

Third, if the international price of a good whose production is pollutive rose because of the imposition of a pollution tax (or import liberalization) abroad, welfare in a small open economy which exports this good would improve—notwithstanding the environmental damage caused by its greater production—so long as a pollution tax close enough to the optimal rate was applied to domestic producers. However, if the economy was an importer of such a good, an international price rise might but need not necessarily make that economy worse off. The country could be better off if it were to switch sufficiently from being an importer to being an exporter of this good and applied a pollution tax close enough to the optimal rate.[14] But even an economy which remains an importer of such a product whose international price has risen could benefit from that rise if (a) domestic production of that product was being discouraged sufficiently by government assistance to other sectors of the economy or (b) other sectors were more pollutive than this sector so that attracting resources to this sector improved the country's environment sufficiently. And conversely for a good whose consumption is pollutive.

The larger the country, the more its activities affect international prices and the larger is its proportionate effect on production, consumption and hence pollution in the rest of the world. This tends to moderate the proportional effects on an economy of liberalizing its trade, although many more exceptional outcomes are possible because of the varying extents to which (a) international environmental spillovers occur, (b) countries differ in their disutility from pollution, (c) imported goods are imperfect substitutes for domestic goods in terms of the consumption externalities they impose, and (d) the rest of the world alters its environmental (and trade) policies in response to policy changes in the country concerned. Even so, the fundamental point remains that free trade is nationally and globally superior to no trade so long as the optimal pollution tax is in place.

The above results depend of course on the assumptions made, so numerous qualifications are needed. Nonetheless, the key results are likely to be strengthened rather than weakened by more comprehensive analyses. For example, relaxing the assumption that the source of pollution is production or consumption *per se*, rather than the process used, ensures that the welfare gains from trade will be greater than suggested above when the optimal environmental policy instrument is used. This is because that optimal instrument will be less costly in terms of resources used or consumption foregone than a production or consumption tax. Relaxing that assumption also increases the cost of using trade policy rather than the optimal environmental policy instrument to overcome environmental externalities. The welfare gains from opening up an economy are also enhanced if by being more open it is less expensive (a) to import or develop domestically new technologies that are environmentally friendlier and/or (b) to import or export mobile factors of production in response to changes in countries' environmental policies. And when many small countries simultaneously impose a pollution tax on production of their exportables, the cost to any one of them of improving their national environment by a given amount is likely to be less, for two reasons: their terms of trade improve in the case of multilateral action; and pollution from abroad is reduced by the other countries' pollution taxes. The only possibility for their environment not to improve is if the increase in production in third (non-taxing) countries adds more degradation than the reduction in emissions in the taxing countries.

An important real-world implication of the above analysis has to do with developing countries' concerns that they will be made worse off as advanced industrial economies tighten their pollution standards. The analysis suggests these concerns are not justified if the advanced countries' imports are relatively pollution-intensive in their production. But if it is the developing countries that continue to import goods whose production is relatively pollutive, they may well have cause for concern. This is because their terms of trade are worsened by the rise in the industrial countries' pollution taxes and their own emissions increase as their import-competing sectors expand. And should developing countries respond by introducing or raising their own pollution taxes, their terms of trade would deteriorate further and thereby limit the extent to which such a response could have an offsetting positive influence of welfare. However, three points should be kept in mind when considering this issue.

First, the more pollution taxes are raised in advanced economies relative to developing economies, the more production of pollution-intensive goods will be relocated to developing countries. This is particularly so if the appropriate capital is internationally mobile, in which case it is more likely that those countries will become exporters of pollutive goods. Once that point is reached, developing countries will be beneficiaries of further pollution tax increases in the advanced economies provided their marginal disutility from a dirtier domestic environment is low or they adjust their own pollution taxes upwards to their new optimal level.

Second, it should not be inferred from the above that poor people do not care for the environment. On the contrary, it is reasonable to assume that rich and poor people have similar tastes and preferences for all goods and services, including those of a clean environment (following Stigler and Becker, 1977). What differs between rich and poor people is their incomes, and hence their capacity and preparedness to trade off greater goods consumption for a cleaner environment. It is therefore not surprising that the (explicit or implicit) rates of taxation of pollution tend to be highly correlated with the per capita incomes of nation states. With this perspective it is clear that pollution abatement is not greatly different from any other item with a high income elasticity of demand, including its effect on the international terms of trade as global incomes rise. There is no more justification for underpricing environmental services as their demand grows, in order not to worsen the terms of trade for a subset of countries, than there is to distort the prices of any other goods or services.

Having said that, the third point to keep in mind is that provided pollution tax rates are not too far from their socially optimal levels in each country, the rich countries whose welfare increases when they raise their pollution taxes need share only some of that gain to compensate fully those poorer countries whose welfare is reduced.

One other important real-world implication of the above analysis relates to the relationship between the demand for a cleaner environment and the pattern of distortions to incentives that is in place. In a distortion-ridden economy, one of the cheapest ways for policy makers to satisfy the increasing demand for less pollution may be not to introduce a pollution tax but to reduce government assistance to pollution-intensive industries. That would be the case if those industries were the most assisted. On the other hand, if pollution-intensive industries were the least assisted by the current pattern of distortions, an economic reform program aimed at reducing distortions may need to be accompanied by the introduction of or an increase in pollution taxes to guarantee that the policy changes improved national welfare.

End Notes

Helpful comments by John Black, Bernard Hoekman, Michael Leidy, Peter Lloyd, Francisco Nadal De Simone, Richard Snape, Arvind Subramanian, John Whalley and Alan Winters are gratefully acknowledged.

1. Where property rights are well defined and transactions costs permit, social costs may be internalized cheapest by the private sector (Coase, 1960). This would be especially so if government intervention is likely to result in the use of second-best policy instruments because of interest-group pressures on politicians, as discussed for example in the chapter in this volume by Hoekman and Leidy.

2. These are standard assumptions in the economic analysis of distortions,

as discussed for example by Corden (1974). While not denying the considerable difficulties associated with measuring the benefits of the environment, as discussed by, for example, Johansson (1987) and Braden and Kolstad (1991), the analysis explicitly assumes that such benefits are not infinite and that costs of pollution abatement are positive.

3. As Krutilla (1991) points out, with production externalities this is the case when emissions are not substitutable with other inputs in the production process, or when reducing output is a lower-cost response to an emissions tax than changing the production process (as it may well be in the short to medium term and/or when low levels of intervention are called for).

4. For a detailed exposition of the appropriateness of this consumer and producer surplus approach to measuring social welfare, see, for example, Just *et al.* (1982).

5. Given the assumptions made, this would bring about what is known in the environmental literature as the Lindahl equilibrium (Mäler, 1985). Note that even when the environment is highly valued it does not follow that the society would be better off the higher the pollution tax, because such a tax reduces environmental degradation at the expense of goods production and hence consumption. In the autarchy case, a production tax of *tg* units, for example, would provide less social welfare than if no environmental policy was in place because *tch* exceeds *cde* in Figure 8.1(a).

6. Further elaboration of the environmental effects of lowering trade barriers is provided in Chapter 8 of this volume.

7. As Britten-Jones *et al.* (1987) point out, the optimal second-best export tax is that which maximizes the difference between those two areas, contrary to the criterion suggested by Corden (1974). Note that a parallel analysis of an import subsidy could be made, but in situations where production taxes are administratively infeasible (notably developing countries), import subsidies are likely to be difficult to implement also because of the high costs of raising the necessary tax revenue.

8. What follows draws to some extent on an analysis (in a different context) by Anderson and Tyers (1991).

9. As Anderson and Tyers (1991) point out, area $cfts = z\Delta P.s$ where ΔP is the change in price, s is the import subsidy per unit and z is the (negative of the) slope of the excess demand curve. Area $fgrj = \Delta P[zs - X] - \frac{1}{2}z\Delta P]$ where X is the quantity that would have been exported if the domestic price had been OP. Hence the difference between these two areas, which is the net welfare gain from the international price rise of ΔP that is fully transmitted domestically ($\Delta P = PP' = P_oP_o'$), is thus $\Delta P(X + \frac{1}{2}z\Delta P)$. This is identical to *adhn* in Figure 8.2. Notice that the above is concerned with the welfare effect of the international and domestic price changes in

the presence of the country's protectionist policy, not with the change in the welfare cost of that protectionist policy.

10. This presumes in particular that if the expansion of agriculture in less-developed economies involved the clearing of forest land, such an expansion would be no worse for the local and global environment than if the resources so used to expand food output were used in other (pollutive) activities. Variations of this example are discussed in more detail in Chapter 8 of this volume.

11. Unlike in the small country case, though, it does not follow from the latter result that liberalizing a less-than-prohibitive trade barrier (as distinct from switching from autarchy to free trade) must improve this large country's welfare, even with the optimal pollution tax in place. This is because the large country by definition has some monopoly power in trade, and the trade tax that it eliminates may be less than (or not sufficiently above) its optimal trade tax from the viewpoint of maximizing its consumption of goods.

12. In what follows the positive externality which the existences of elephants bestows on Africa's tourist industry and the negative externality which roaming elephants impose on African farmers and villagers are assumed exactly to offset each other.

13. Here only ivory tax policy instruments are considered. There is, however, the possibility of better national park management policies and the like being able to boost elephant herds sufficiently at even lower cost.

14. This possibility of a switch in an importing country's trade pattern raises the more general question of what determines changes in a country's comparative advantage in pollutive goods in a growing world economy, an issue that is addressed in, for example, Blackhurst (1977).

References

Anderson, K. and R. Tyers (1991), 'More on Welfare Gains to Developing Countries from Liberalizing World Food Trade', mimeo, Australian National University, Canberra, July.

Barbier, E., J. Burgess, T. Swanson and D. Pearce (1990), *Elephants, Economics and Ivory,* London: Earthscan.

Blackhurst, R. (1977), 'International Trade and Domestic Environmental Policies in a Growing World Economy', pp. 341–64 in *International Relations in a Changing World,* by R. Blackhurst *et al.,* Geneva: Sythoff-Leiden.

Braden, J. B. and C. D. Kolstad (eds.) (1991), *Measuring the Demand for Environmental Quality,* Amsterdam: North Holland.

Britten-Jones, M., R. S. Nettle and K. Anderson (1987), 'On Optimal Second-Best Trade Intervention in the Presence of a Domestic Divergence', *Australian Economic Papers* 26: 332–36.

Coase, R. H. (1960), 'The Problem of Social Cost', *Journal of Law and Economics* **3**: 1–44.

Corden, W. M. (1974), *Trade Policy and Economic Welfare,* Oxford: Clarendon Press.

Johansson, P-O. (1987), *The Economic Theory and Measurement of Environmental Benefits,* Cambridge: Cambridge University Press.

Just, R. E., D. L. Hueth and A. Schmitz (1982), *Applied Welfare Economics and Public Policy,* Englewood Cliffs: Prentice Hall.

Krutilla, K. (1991), 'Environmental Regulation in an Open Economy', *Journal of Environmental Economics and Management* **20**: 127–42.

Mäler, K.-G. (1985), 'Welfare Economics and the Environment', Chapter 1 in *Handbook of Natural Resources and Energy Economics*, Vol. 1, edited by A. V. Kneese and J. L. Sweeney, Amsterdam: North Holland.

Newbery, D. M. (1990), 'Acid Rain', *Economic Policy* **5**, (11), 297–346.

Stigler, G. J. and G. S. Becker (1977), 'De Gustibus Non Est Disputandum', *American Economic Review* **67**: 76–90.

9

General Models of Environmental Policy and Foreign Trade

Karl W. Steininger

9.1 INTRODUCTION

The path towards answering environmental policy and foreign trade questions lies between—and draws upon—two subfields of economics, namely international and environmental economics. Both are very rich areas in their own right, and a survey of their interlinkage cannot but be selective. This chapter seeks to provide a link between a categorization of the issues on the research agenda, and the body of general equilibrium theory as well as a link to empirical research to shed light on these areas. Some of the links have already been explored; others point to areas of future work. Interest in the field has increased dramatically since the early 1990s. The aim of this survey is to give the flavour of a subject which remains a vibrant and fruitful source of theoretical insight, of testable hypotheses and of illuminating quantification.

The rapid integration of the world economy puts questions such as the following high on the agenda:

- Which countries export goods the production of which is pollution-intensive?
- Can a country unilaterally introduce an energy or CO_2 tax without significant implications for its trade flows or for industry migration?
- How does trade liberalization affect the environment?

Each of these questions represents one key area in a threefold classification of trade and environment issues within economics: (a) in terms of positive theory, the objective of international economics is to explain the manifestation of the international division of labour and resulting trade flows (the pattern of trade). This increasingly needs to take into account factor endowment and use of environmental and natural resources. Further, the subject of (positive and normative) analysis is policy evaluation for the following linkages: (b) the effect of environmental policy on trade flows and (c) the effect of trade policy on the state of the environment.

We can ask which modelling approach is the appropriate one, or, more specifically for this chapter, when do we need a general model? With respect to (a), explaining the patterns of trade is a task which by its very nature requires a general model, that is, a model covering the full range of products (the exported goods and the imported goods they are exchanged for) and all factors of production available within a country. Partial models as covered are not sufficient to determine in what range of produced goods a country specializes. With respect to (b) and (c), that is, in the area of policy evaluation, the specific nature of the question asked determines whether a partial or a general model is appropriate. A general model covers the interlinkages among sectors, between agents' incomes and expenditures, among factor markets and thus also the feedback effects of policy. The use of a general model is required where the following aspects are concerned:

- *type of policy* examined: when trade liberalization concerns either the whole economy or crucial sectors of it, or when environmental policy produces economy-wide impacts, as is the case with energy policy or CO_2 policy;
- *type of effects* analysed: when the distribution of impacts across agents within one country (under any grouping), or when the impact distribution among countries is of concern;
- *importance of net balances:* when the structure of the balance of payments (for example a balanced current account) is of relevance, or, on the environmental side, the net (for example, global) environmental impact is of importance (leakage effect).

The structure of this chapter is as follows. Section 2 focuses on the underlying theoretical basis. This has developed primarily as environmental extensions to the theory of international trade. Empirical evaluation is surveyed thereafter, split into past experience (section 3) and simulation analysis (section 4). The outlook for future research concludes the chapter.

9.2   ENVIRONMENTAL EXTENSIONS IN TRADE THEORY

9.2.1 Environmental Application of Standard Trade Theorems

A natural first step in analysing the trade and environment interlinkage is to expand the existing body of trade theory by including environmental considerations. The foundation for explaining trade patterns laid by Ricardo focuses on the notion of comparative cost advantage as opposed to absolute cost advantage. The Ricardian model identifies (i) differences in the productivity of one factor (labour) as the source of differences in relative pre-trade (costs and) prices, and (ii) the comparative pre-trade prices as explaining trade patterns. As far as environmental questions are concerned, with respect to (i), above environmental factors are of insignificant importance in explaining intercountry differences in labour productivity. However, with respect to (ii), which became the core of international economics, environmental expansion of its multi-factor representation, the Heckscher–Ohlin (H–O) model, is more fruitful.

The H–O model (for an exposition see, for example, Bhagwati and Srinivasan, 1983, chs 5–7) is an application of neoclassical general equilibrium theory to international economics. It focuses on differences between countries in their relative factor endowments, and generates results primarily in three areas:

1. *Patterns of trade (Heckscher–Ohlin theorem):* A country will specialize in the production (and export) of the good intensive in that factor which is in abundant supply at home, which for a country with high assimilative environmental capacity, for example, will be the pollution-intensive good. The introduction of an environmental policy (such as a tax) will cause a shift toward or–given a sufficient level–even establish a comparative advantage in the 'clean' good (Pethig, 1976; Siebert *et al.*, 1980; Siebert, 1995, p. 171 ff.). The H–O model, however, rests on quite restrictive assumptions, for example, same technology across countries, constant returns to scale, only two goods and two factors. For the multi-good, multi-factor case a weaker version of the patterns-of-trade conclusion holds in terms of factor content of goods: a country tends to import those factors that are relatively expensive pre-trade (Heckscher–Ohlin–Vanek (H–O–V) theorem). For local and regional environmental resources, for which the assumption of international factor immobility is highly valid, this implies that the higher the pre-trade resource price relative to prices of additional production factors, the

more this resource will be imported, in the form of goods whose production is intensive in these resources.

2. *Factor prices (factor-price equalization theorem):* Free trade is shown to equalize factor prices such that trade in goods (or international mobility of some factors) substitutes for migration of immobile factors (for example labour, environment). It implies that environmental shadow prices (environmental taxes) will equalize across countries (Siebert, 1987, p. 169 and 1995, p. 177). Intuitively, the country with less assimilative capacity will focus on the production of goods which are cleaner and thus raise pollution abroad until shadow prices are equal. Note that this does not imply the same environmental quality in the trading partner countries. However, this result does not hold whenever mobility of the factor labour depends on environmental quality. Labour will then migrate to the cleaner area and further increase the demand for environmental goods in that area, implying higher pollution taxes (and a segmented labour market) (Siebert as cited).

As with the traditional version focusing on wage equalization, the importance of the theorem lies in its ability to explain the sources of real world *deviation* from environmental factor-price equalization as each of its rather rigid assumptions are lifted (cf. Haberler, 1961, p. 18). Deviations occur.

• with increasing rather than constant returns to scale (country size differences become decisive–Markusen and Melvin, 1981);

• with imperfect rather than perfect competition (bargaining power of factor owners becomes the relevant factor, for a survey of imperfect competition models see Jones and Neary, 1984, p. 50ff. and Krugman, 1995);

• with internationally non-identical production functions (differences in technological progress across industries and/or countries become decisive); and

• with complete specialization (factor endowment becomes decisive, for example Bhagwati and Srinivasan, 1983, p. 61ff.).

3. *Income distribution (Stolper-Samuelson theorem):* The protection of an industry, if it raises the domestic goods price, will cause the owners of the factor intensively used in producing that good to gain absolutely. When environmental property rights are allocated in a country and the use of the factor environment is rewarded, protection of environmentally intensive industries

thus benefits the owners of environmental resources absolutely. However, in general, for the 'more than two goods and more than two factors case' only a weaker version holds true: a given (policy-induced) range of changes in commodity prices gives rise to a larger range of changes in factor prices (what Jones, 1965 has called the 'magnification effect'), implying only that 'there is some factor' whose real return goes up, but not necessarily the one intensively used (see Ethier, 1984, p. 165).

Summarizing, the application of the standard trade theorems in models including environmental and natural resource factors does allow us to advance to some extent. While partial equilibrium analysis of environmental policy only allows welfare conclusions to be made once knowledge is available on whether the good prior to policy is imported or exported (for a comprehensive exposition see Anderson, 1992), the H–O model specifies exactly this prerequisite knowledge, that is, which country exports and which country imports the environment-intensive good. Further, this approach also identifies why environmental shadow prices (for example taken as environmental taxes) do not equalize across countries. Also, the international income distribution effects of environmental policy can be derived.

However, the multitude of interdependences within a general equilibrium model implies that in order to reach definite conclusions quite restrictive assumptions are required. The H–O model itself, focusing on factor endowment differences, is considered a particular case of the neoclassical general equilibrium model in which internationally identical production functions and tastes are assumed. Neoclassical trade theory beyond the H–O approach finds the determinants of trade simultaneously in three areas: in the differences between technologies, factor endowments and tastes of different countries.[1] Considering such simultaneous interactions entails an extremely high degree of complexity and often prevents clearcut analytical solutions. Numerical solutions have thus increasingly gained importance, with the results of the class of computable general equilibrium simulations being covered here in Section 4. Theory deals with this complexity in a different way. Beyond the environmental results of the standard trade theorems stated above, the contribution of theory is to alter a (rather basic) H–O model regarding generally just one aspect at a time, which is considered the most relevant for a particular trade and environment question.

9.2.2 Issue-specific Environmental Extensions to Trade Theory

This line of research turned out to focus on the environmental effects of international economic liberalization. The approaches differ on two grounds: (a) the aspect of liberalization and (b) the primary causality determining environmental quality.

Rauscher (1991) abandons the assumption of factor immobility of capital in a model with no goods trade, and Rauscher (1995, 1997) additionally includes consumption externalities (pollution) and external effects on production (production is influenced by environmental quality). He finds that increased capital mobility (increased 'openness') leads to pollution reduction in at least one country, but has ambiguous welfare effects, especially for the large-country case, due to transfrontier pollution spillover. Overall, international environmental problems may be aggravated by international capital mobility.

Chichilnisky (1994) allows for endogenous factor supply (namely of the environmental resource), governed by the degree of establishment of a system of property rights. The 'North' has a well-established system, in contrast to the 'South', where resources are subject to open access. International trade simply increases the problem of over exploitation, with the definition of private property rights as the only feasible policy response in the model.[2]

Copeland and Taylor (1994) endogenize the level of environmental policy. They allow for intercountry differences in the endowment of effective labour (human capital), which results in income differences. The demand for the environment, which is taken to be a normal good, increases with income. Considering only local pollution, they find that free trade shifts pollution-intensive production to the country where human capital is relatively scarce (the poor South) and raises world pollution. The impact in any one country can be separated into the scale effect (increasing pollution), the composition effect (reflecting different international specialization in more or less dirty products) and the technology effect (indicating the move towards cleaner technology).

Copeland and Taylor (1995) enlarge the analysis to cover the global commons (as well as the multi-country case). They find that if world distribution of income is highly skewed, free trade will harm the environment, but also that international trade in emission permits can reduce this negative impact, that lower-income countries have an incentive to delay international pollution agreements until after multilateral trade liberalization has been achieved, and that income transfers—only if tied directly to pollution reduction–can be welfare-enhancing.

9.3 ANALYSING PAST EXPERIENCE: STATISTICAL AND ECONOMETRIC APPROACHES

Looking at empirical evidence, statistical and econometric approaches have been used to analyse quantitatively the question of whether environmental policy has influenced trade flows and industry location. This is done to determine the significance of political concerns which vary according to the environmental policy instrument chosen. The use of environmental *economic instruments* (taxes, permits, subsidies) raises *domestic* concerns regarding industry competitiveness, whereas *product or production process regulation* (binding or voluntary, for example eco-labelling) raises *foreign* concerns with respect to market access.

Ideally, one would like to test empirically the effects of environmental policy on competitiveness by identifying the effect that the policy would have on net exports, holding real wages and exchange rates constant. In an open economy, however, the fall in net exports in one industry will be balanced quickly by the rise of net exports in another industry, an industry which, if the change is solely brought about by exchange rate changes (or a fall in real wages) should not be thought of as having gained competitiveness.[3] Noting that this ideal form of the empirical text is essentially impossible to implement in practice, Jaffe *et al.* (1995) categorize the three alternative indicators of competitiveness used in empirical studies as follows:

  (i) changes in net exports of heavily regulated industries relative to the net exports in more lightly regulated industries

 (ii) changes in the share of world production in heavily regulated industries indicating shifts in the loci of production

(iii) changes in the geographic destination of foreign direct investment flows of heavily regulated industries.[4]

Most research to date has focused on the first indicator, with two surveys available: Ugelow (1982) reviews the studies of the 1970s and Dean (1992) those up to 1990. The broad overall finding was that the effects of environmental regulation on trade patterns have been rather small, where noticeable at all. Dean (1992, p. 16) concludes:

> The methodologies are quite varied, making comparisons between studies difficult. However, some generalizations can be drawn. First, estimates of total environmental control costs (ECC) by industry

tend to be very low–abatement costs are a very small portion of industry costs on average. Second, reductions in output caused by ECC are also small and insignificant on an average, though they can be significant for some individual sectors. Third, there is little evidence of any significant impact of ECC on the pattern of trade.

Environmental control cost data can be used in two types of analyses. For quantifying direct ECC effects, time-series and/or cross-industry data are used in a regression framework. Control costs, however, can be passed on to intermediate input users further along in the production sequence. To cover both direct and indirect ECCs (price increases of intermediate inputs) a model of an input–output basis must be employed. With a general equilibrium model being the most comprehensive representative of this second type, the incidence of upstream costs changes can be determined to a degree sufficient for most analyses by combining the input–output table with direct ECC (Kalt, 1988). While the time-series/cross-industry approach neglects indirect costs, it is applicable at a very disaggregated industry level. For the second approach, which includes indirect costs, the effort involved in constructing the input–output table limits the possible degree of disaggregation.

That the trade impact of direct and indirect ECC data is not significant is reconfirmed by the studies of the early 1990s. Grossman and Krueger (1993) employ a regression framework and, even though they focus on US-Mexico trade, and thus on two countries that experience both a large volume of trade due to geographical proximity and significant historical differences in environmental laws, they find that the coefficient of pollution abatement control cost is not significant in explaining US imports. They regress imports on factor shares, the effective tariff rate and pollution abatement control costs. Tobey (1990, 1993), focusing on 24 pollution-intensive agricultural and manufacturing industries and data from 23 countries in a cross-section H–O–V model, similarly does not find any significant influence of environmental regulation stringency on trade flows.

The central limitation of these studies can be seen in the representation of environmental policy stringency. First, there are no internationally comparable ECC data available. Second, data from firm surveys are likely to under-report ECCs if respondents do not have full knowledge of all costs. Also, determination of costs is increasingly difficult, the more integrated environmental technology becomes. Finally, ECCs need to be related correctly, that is, direct ECCs are to be expressed as a ratio of value added, and direct plus indirect ECCs as a ratio of total costs (Kalt, 1988). Tobey's analysis, for example, has been criticized by the Office of Technology Assessment (1992, p. 101) on the second count. For a detailed survey on the studies and their limitations see Steininger (1995, pp. 82–92).

When van Beers and van den Bergh (1997) eliminate both the above-mentioned as well as a further potential drawback of Tobey's analysis by selecting the environmental policy indicator better to reflect private environmental production costs, and by focusing on bilateral instead of multilateral trade flows, they do find significant export effects of environmental policy on total trade flows and on 'dirty' industries' trade flows. For the latter, however, their results hold only for the subgroup of footloose (non-resource-based) industries. For imports the impact is negative as well, that is, import barriers may have been installed, together with stricter environmental regulations.

With respect to the second indicator of competitiveness, that is, shifts in the locus of production of pollution-intensive goods, Low (1992) contains a number of empirical studies, while the standard comprehensive earlier reference is Leonard (1988). A number of studies do find evidence that pollution-intensive industries have shifted to low-income countries in the South (for example Low and Yeats, 1992, Lucas et al., 1992, Birdsall and Wheeler, 1992). This result is open to interpretation, however. Beghin et al. (1994) consider the contrast between the findings that the South specializes in dirty industries and that environmental regulations have only modest effects on competitiveness to signal measurement problems (under-reporting of environmental costs). Jaffe et al. (1995), in their survey of the empirical evidence, regard this firm migration as being small, when seen in the overall context of economic development (for example the share of pollution-intensive products in Southeast Asia rising from 3.4 to 8.4 per cent in the period 1965–88 (Low and Yeats, 1992), while the lion's share of the world's exports of 'dirty' products is still accounted for by industrialized countries). Also, the move of production may simply be due to increased demand in those countries. Finally, natural resource endowments significantly contribute to explaining the patterns of pollution-intensive exports. They conclude that it is 'by no means clear that the changes in trade patterns were caused by increasingly strict environmental regulations in developed countries. (The observed changes). . . are consistent with the general process of development in the Third World.'

Where environmental regulation can indeed play a significant role causing location shifts in production, is in individual (sub)sectors of the economy. Several case studies do exist; for example Lesperance (1991) studied the wood product coatings industry, which is heavily regulated in California, and found that employment shifted to Mexico.

With respect to the third indicator, foreign direct investment, the impact of environmental regulation on complex investment decisions is even more difficult to isolate. Of the very few studies available, Leonard (1988) documents that there is no systematic pattern of foreign direct investment in polluting industries. Also, more recent empirical work has not been able to establish a systematic pattern that is beyond doubt.[5] This finding may reflect the fact that weak environmental standards often go hand in hand with other factors that deter investment, such as political instability, uncertainty about future regulation, and corruption. Such factors are of equal or greater importance than low environmental regulation costs.

With respect to all three classes of studies, one criticism mentioned above and relevant for many other empirical evaluations is the determination of environmental control costs. The backward-looking perspective of these studies must also be considered a limiting factor as the importance of environmental control costs is likely to increase in the future (OTA, 1992, p. 97). While in such backward-looking analysis the first limitation is difficult and the second impossible to overcome, they can both be dealt with in prospective simulation studies.

9.4 SIMULATION ANALYSIS: THE COMPUTABLE GENERAL EQUILIBRIUM (CGE) APPROACH

Computable general equilibrium modelling provides a suitable tool for empirical analysis when it is the general, economy-wide effects of policy which are of concern. CGE modelling is an approach that builds upon one of the most fundamental ideas of economics, namely that of grasping the complex interdependence among the different markets in an economy by taking the outcome to represent a 'general equilibrium'. Dating back to Walras, it was formally analysed by Arrow and Debreu, and has subsequently been extended–parallel to the development of solution algorithms–for practical empirical policy analysis. A CGE model contains explicit modelling of (i) the behaviour of individual agents (households, producers), (ii) market clearing of factor and goods markets, (iii) budget constraints for agents, institutions, as well as macroeconomic balances (based on the Social Accounting Matrix), and (iv) sectoral interlinkages in production (based on the input–output table).

Two of the main classes of applications for CGE models are international trade on the one hand and energy and environment on the other. The interlinkage of the two fields has also been seen to lend itself naturally to this type of analysis. In addition to the modelling characteristics relevant for environmental CGE models (such as description of taxes/ permits, abatement technology, and revenue recycling modelling), for the trade and environment interlinkage the type of trade flow modelling is crucial.

It is the assumption on the degree of substitutability between imported and domestic goods that most importantly govern the range of the quantitative environment and trade results. CGE modelling originally started with the traditional trade theory assumption (Heckscher–Ohlin) of perfect interchangeability of imported and domestic goods. Two important real-world trade features are inconsistent with such modelling: two-way trade and pervasive violation of the law of one price. Modellers reacted with a multi-stage specification of the demand structure. In the first stage, total expenditure is allocated between all the production sectors. In the second stage, expenditure for each industrial sector is allocated between imports and competing domestic production. The most widely used form for this second stage is *national* product differentiation combined with the constant elasticity of substitution (CES) functional form ('Armington assumption', Armington, 1969). This delinks the domestic and international price level and avoids complete specialization. Generally, however, elasticities of substitution between varieties of goods will vary over time and in addition be unequal between pairs (country A and B; country A and C) of varieties (on the latter see Winters, 1984). The specification of the second stage has thus been generalized to account for these two features by using flexible functional forms for the demand specification, such as AIDS (almost ideal demand system). An alternative, chosen by other modellers, is to switch to firm-level, rather than national product differentiation. Finally, one can also return to perfect substitutability between domestic and foreign products, at least for some sectors modelled (small open economy). In this case, with constant returns to scale, however, there will only be as many goods produced domestically as there are primary factors (Samuelson, 1953); if there are more goods, complete specialization will occur. Complete specialization can also be avoided by limiting sectoral factor mobility, by introducing sectorally specific capital vintages subject to an explicit depreciation process (as in Bergman, 1991).

The reaction of (sectoral) domestic production to environmental policy in general is largest in models that assume high or perfect substitutability between domestic and foreign (varieties of) products. The same holds for the reverse policy link: trade liberalization moves domestic prices closer to the international price level the higher international competition in the sector is, with the larger quantity adjustments triggered implying a stronger environmental quality change. Many of the CGE models on environment and trade seek to establish upper bounds of potential impacts and thus include at least some sectors modelled as price-taking. Usually the sectors best characterized as price takers are also those of highest pollution intensity (basic industries).

9.2.3 CGE Modelling Conclusions

A major concern with the introduction of environmental policy is the competitive position of exposed industries in small open economies. Bergman (1991), in a seven-sector model of the Swedish economy, distinguishes two exposed sectors as price takers. For rather significant emission reduction targets (in a 15-year period), compared with unconstrained development, emissions of SO_x are reduced by 44 per cent, NO_x by 57 per cent and CO_2 by 28 per cent, and while the aggregate impacts on production are comparatively small, the output in one of the exposed sectors (Steel and Chemicals) is cut to below half. As mentioned above, this model distinguishes sectoral capital vintages. Assumptions about the depreciation rate and technological change in emission intensity relevant for these vintages have a strong influence on the result. In a 19-sector Armington specification for Austria designed to estimate the cost of meeting the Toronto objective, Breuss and Steininger (1998) find a similar result for the sector Base Metals, with the sectors Paper and Wood, Mining, and Petroleum following successively in output reduction. They also show that choosing sectoral differentiation in emission-revenue recycling as a means of temporarily slowing sectoral adjustment can strongly mitigate sector-specific production losses, although this is achieved at the cost of up to 50 per cent higher environmental tax rates. While there is more long-term experience with environmental policy in the US, sectoral foreign trade concerns are not at the centre of analysis for the large-country case (for example Boyd and Krutilla, 1992; Jorgenson and Wilcoxen, 1995).

How strongly do the costs of environmental policy differ for unilateral and multilateral policy implementation? Multi-country modelling provides the appropriate framework for answering this question and has primarily been applied to issues of the global commons (the greenhouse effect). As a yardstick for measuring and comparing costs we are used to GDP figures. Such figures are, however, only a partial indicator of welfare, as they fail to take into account, *inter alia*, changes in the terms of trade and the consumption losses due to environmental taxes (as noted by, for example, Dean and Hoeller, 1992). Most recent studies thus prefer to use real income changes. Representing the results in terms of equivalent variations would be more revealing, but is not yet common in multi-country modelling. The OECD GREEN (General Equilibrium Environment) model, distinguishing 12 world regions, has been employed to analyse unilateral versus coordinated greenhouse policy (Burniaux *et al.*, 1992). It was found, for example, that unilateral OECD emission stabilization comes at a real income loss of 0.6 per cent for the period 1995–2050, while a worldwide Toronto-type agreement implies a sixfold reduction of global emissions and causes a 1.2 per cent real income loss for the OECD. When this higher worldwide emission reduction is achieved in cost-effective terms (equalizing marginal costs across regions) the real income loss for the OECD is reduced to 0.5 per cent.

Similarly, when the EC acts unilaterally within the OECD, costs are above the case of overall OECD action, but only slightly. When comparing unilateral German greenhouse policy action with EC-wide action in a three-region model, Welsch and Hoster (1995) find that the impacts on German output are much *smaller* for unilateral action, due to German dependence on export to other EC countries.

For the distribution of welfare impacts across countries, CGE models indicate that it is not only the choice of the base of the tax (for example production or consumption) that is important (Piggot and Whalley, 1992), but also that terms-of-trade effects are significant. In the context of greenhouse policy this is particularly true for oil exporters and oil-importing LDCs (GREEN model, Burniaux *et al.*, 1992).

Considering policy effectiveness, a central issue for *global* environmental problems is leakage, that is, unilateral emission reduction increasing emissions abroad due to changed trade structures. For unilateral greenhouse policy, leakage can occur through three channels of international trade: (i) emission-intensive manufacturing relocating abroad; (ii) demand-induced reduction of world oil prices increasing oil use abroad; and (iii) substitution of oil by natural gas which is less greenhouse-effective and induces the reverse substitution effect aboard. [Manne (1994) identifies the energy-intensive manufacturing link as the largest, and the gas link as substantial, but primarily in the medium term]. The quantification, however, is crucially dependent on the foreign trade modelling chosen. The leakage rates resulting from 'Armington' models of imperfect goods substitutability (Burniaux *et al.*, 1992) are much smaller than those derived from models that use the Heckscher-Ohlin assumption of perfect substitutability (Rutherford, 1992; Felder and Rutherford, 1993). Manne (1994) offers an intermediate approach by allowing for perfect substitutability but imposing quadratic penalties for deviations from base-year trade flows, thus moderating the Armington assumption. For a unilateral OECD 20 per cent emission cutback he quantifies leakage rates, that is, foreign emission increase as a percentage of domestic reduction, of up to 30 per cent. However, the long-term potential is higher: when Manne allows for energy-intensive production to be completely phased out in the OECD by the time today's workers retire, imports of energy-intensive production rise fivefold, implying a substantial leakage increase.

9.5   CURRENT TRENDS AND PROSPECTS

9.5.1 Space Matters

Environmental and natural resource considerations offer a strong potential for supplying missing building blocks in trade-theory development. The significance of geographical distance in trade theory and policy has been largely neglected since Marshall. Its importance has only recently been pointed out again by Krugman (for example, Krugman, 1991). Within this new development in trade theory, geography and trade, there is still some "'mystery'" ... [left on] the origins of economically meaningful regions'. The role of environmental services could help to explain these origins (Smith and Espinosa, 1996). Environmental sciences emphasize the spatial dimension and interactions of environmental impacts in physical terms (acknowledged in a trade and environment analysis by Perroni and Wigle, 1994). Also the demand for goods is found to be dependent on environmental preferences present in a region. Regions can thus be defined based on simultaneous acknowledgement of natural science and social science dimensions.

9.5.2 Environmental Innovation

Innovation is acknowledged as a major force in economic development at both the firm level (industrial organization) and at the aggregate level (endogenous growth). Hall (1994) and Kwaśnicki (1996) give recent surveys of the field; the collection by Carraro (1994) links innovation to trade and environment for both positive analysis and policy design. It focuses on policy coordination for solving global commons issues. However, several questions remain open. What market and social forces beyond fiscal policies enhance the environmental efficiency of innovations? What market structures and policy approaches best foster their international distribution? Adequate answers to these and related questions have yet to be found. In both theoretical and empirical terms this area remains an ample field for future research.

In the modelling of global environmental problems, developing countries represent one or more country blocks within both theoretical and numerical approaches, as covered above. These analyses leave unanswered, however, the extent to which changes in trade and/or environmental policy in an open economy affect local environmental problems. We know that open trade systems magnify the effect of environmental externalities (see Munasinghe and Cruz, 1995), but how large these effects are, and how policies interlink, must remain the subject of further analysis. The CGE model by Munasinghe and Persson (1995) for analysing deforestation in Costa Rica indicates that quite distinct modelling features are required, for example the introduction of property rights in a numerical model formulation and the interlinking of the markets for logs and cleared land.

Overall it can be said that the building of general models of trade and the environment, along the lines of the respective subfields, benefits greatly from findings in these areas, but in the future is itself likely to be a source of feedback in turn, that is, it may become a source of conceptual and methodological input for both the theoretical and empirical literature. Aside from analytical rigour, however, in a world exhibiting such a rapid pace of economic integration and environmental change as ours, it is to be hoped that researchers will not only select the most crucial issues, but, what is of equal importance, allow their results to be translated into the world of politics.

End Notes

1. The first two areas are touched on in this chapter. Linder (1961) should be consulted for details of the last area.

2. While Chichilnisky (1994) relies on positive factor supply elasticity, van Beers and van den Bergh (1996) further point out technological differences and production externalities as alternative means to introduce environmental considerations by changing the production function assumptions of the H–O model.

3. The term competitiveness in the environment-policy context mainly focuses on competitiveness of particular industries, which is related to but different from competitiveness of a country as a whole (for a trenchant criticism on common but misguided statements on the latter see Krugman, 1994). The concern with the loss of individual industry competitiveness from a national perspective is fourfold (see Jaffe et al., 1995): trade-imbalance-caused devaluations lead to a deterioration in the terms of trade and thus in the domestic standard

of living; the loss is likely to concern specific industries of mainly low-skilled labour with distributional consequences; it may concern sectors that are important for national security; and finally, the adjustment process of sectoral shifts will be connected to a broad set of social costs—irrespective of the final international division of labour which represents an equilibrium again.

4. Beyond these relationships Jaffe *et al.* also point out a fourth and more fundamental (especially long-term) competitiveness link: environmental policy and productivity. See the chapters on growth and the environment in Part VI for a detailed discussion of this issue.

5. In a draft paper, Xing and Kolstad (1995) find that SO₂ emissions (used as an indicator of environmental laxness) significantly explain inflows of foreign direct investment. The causality, however, may also be running in the opposite direction, from investment to pollution (Jaffe *et al.*, 1995). For strictly regulated Germany, the OECD (1993, p. 112) concludes that 'environmental policy does not deteriorate its attractiveness as a location for industry to any decisive degree', while Bouman (1998) finds some evidence that German outward direct investment is indeed affected by domestic environmental costs, though, as he also states, 'the results are not always unambiguous'.

References

Anderson, K. (1992), 'The standard welfare economics of policies affecting trade and the environment', in K. Anderson and R. Blackhurst (eds), *The Greening of World Trade Issues*, New York: Harvester Wheatsheaf.

Armington, P. (1969), 'A theory of demand for products distinguished by place of production', *International Monetary Fund Staff Papers 16*, Washington: IMF, pp. 159–78.

Beers, C. van and J.C.J.M. van den Bergh (1996), 'An overview of methodological approaches in the analysis of trade and environment', *Journal of World Trade*, **30**(1), 143–67.

Beers, C. van and J.C.J.M. van den Bergh (1997), 'An empirical multi-country analysis of the impact of environmental regulations on foreign trade flows', *Kyklos*, **50**, 29–46.

Beghin, J., D. Roland-Holst and D. van der Mensbrugghe (1994), 'A survey of the trade and environment nexus: global dimensions', *OECD Economic Studies*, **23**, Winter, Paris: OECD, pp. 167–92.

Bergman, L. (1991), 'General equilibrium effects of environmental policy: a CGE-modeling approach', *Environmental and Resource Economics*, **1**, 43–61.

Bhagwati, J.N. and T.N. Srinivasan (1983), *Lectures on International Trade*, Cambridge, MA: MIT Press.

Birdsall, N. and D. Wheeler (1992), 'Trade policy and industrial pollution in Latin America: where are the pollution havens?', in Low (1992), pp.

159–67.

Bouman, M. (1998), *Environmental Costs and Capital Flight*, Tinbergen Institute Research Series 177, University of Amsterdam: Thesis Publishers.

Boyd, R. and K. Krutilla (1992), 'Controlling acid depositions: a general equilibrium assessment', *Environmental and Resource Economics*, 2, 307–22.

Breuss, F. and K. Steininger (1998), 'Biomass energy use to reduce climate change: a general equilibrium analysis for Austria', *Journal of Policy* or *Modelling*, **20**(4), 513–35.

Burniaux, J.-M., J.P. Martin, G. Nicoletti and J. Oliveira Martins (1992), 'The costs of reducing CO_2 emissions: evidence from Green', Economics Department Working Paper 115, Paris: OECD.

Carraro, C. (1994), *Trade, Innovation, Environment*, Dordrecht: Kluwer Academic Publishers.

Chichilnisky, G. (1994), 'North–South trade and the global environment', *American Economic Review*, **84**, 851–75.

Copeland, B.R. and M.S. Taylor (1994), 'North–South trade and the environment', *Quarterly Journal of Economics*, **109**, 755–87.

Copeland, B.R. and M.S. Taylor (1995), 'Trade and transboundary pollution', *American Economic Review*, **85**, 716–37.

Dean, J. (1992), 'Trade and the environment: a survey of the literature', in Low (1992), pp. 15–28.

Dean, A. and P. Hoeller (1992), 'Costs of reducing CO_2 emissions: evidence from six global models', Economics Department Working Paper 122, Paris: OECD.

Ethier, W.J. (1984), 'Higher dimensional issues in trade theory', in R.W. Jones and P.B. Kenen (eds), *Handbook of International Economics* Vol. 1, Amsterdam: North-Holland, ch. 3.

Felder, S. and T. Rutherford (1993), 'Unilateral CO_2 reductions and carbon leakage: the consequences of international trade in oil and basic materials', *Journal of Environmental Economics and Management*, **25**, 162–76.

Grossman, G.M. and A.B. Krueger (1993), 'Environmental impacts of a North American Free Trade Agreement', in P. Garber (ed.), *The U.S.–Mexico Free Trade Agreement*, Cambridge, MA: MIT Press, pp. 13–56.

Haberler, G. (1961), 'A survey of international trade theory', *Special Papers in International Economics* 1, International Finance Section, Department of Economics, Princeton University, 78pp.

Hall, P. (1994), *Innovation, Economics and Evolution*, New York: Harvester Wheatsheaf.

Jaffe, A.B., S.R. Peterson, P.R. Portney and R. Stavins (1995). 'Environmental regulation and the competitiveness of U.S. manufacturing', *Journal of Economic Literature*, **33**, 132–63.

Jones, R.W. (1965), 'The structure of simple general equilibrium models', *Journal of Political Economy*, **73**, 557–72.

Jones, R.W. and J.P. Neary (1984), 'Positive theory of international trade' in R.W. Jones and P.B. Kenen (eds), *Handbook of International Economics*,

Vol. 1, Amsterdam: North-Holland, ch. 1.

Jorgenson, D.W. and P.J. Wilcoxen (1995), 'Intertemporal equilibrium modelling of energy and environmental policies' in P.-O. Johansson, B. Kriström and K.-G. Mäler (eds), *Current issues in environmental economics*, Manchester: Manchester University Press.

Kalt, J. (1988), 'The impact of domestic environmental regulatory policies on U.S. international competitiveness', in A.M. Spence and H.A. Hazard (eds), *International Competitiveness*, Cambridge, MA: Ballinger, pp. 221–62.

Krugman, P. (1991), *Geography and Trade*, Cambridge, MA: MIT Press.

Krugman, P. (1994), 'Competitiveness: a dangerous obsession', *Foreign Affairs*, **73**, March/April, 28–44.

Krugman, P. (1995), 'Increasing returns, imperfect competition and the positive theory' in G.M. Grossman and K. Rogoff (eds), *Handbook of International Economics*, Vol. 3, Amsterdam: North-Holland, ch. 24.

Kwašnicki, W. (1996), *Knowledge, Innovation and Economy*, Cheltenham, UK and Brookfield, US: Edward Elgar.

Leonard, J. (1988), *Pollution and the Struggle for World Product: Multinational Corporations, Environment and International Comparative Advantage*, Cambridge, UK: Cambridge University Press.

Lesperance, A.M. (1991), 'Air quality regulations and their impacts on industrial growth in California: a case study of the South Coast Air Quality Management District Rule 1136 and the wood products coatings-industry, Master's thesis, University of California, Los Angeles.

Linder Burenstam, S. (1961), *An Essay on Trade and Transformation*, Upsala: Almquist and Wiksell, and New York: John Wiley and Sons.

Low, P. (ed.) (1992), *International Trade and the Environment*, World Bank Discussion Paper 159, Washington, DC: World Bank.

Low, P. and A. Yeats (1992), 'Do dirty industries migrate?', in Low (1992), pp. 89–103.

Lucas, R.E., D. Wheeler and H. Hettige (1992), 'Economic development, environmental regulation and the international migration of toxic industrial pollution: 1960–1988', in Low (1992), pp. 67–86.

Manne, A.S. (1994), 'International trade: the impact of unilateral carbon emission limits', in OECD, *The Economics of Climate Change*, Paris: OECD, pp. 193–205.

Markusen, J.R. and J.R. Melvin (1981), 'Trade, factor prices and the gains from trade with increasing returns to scale', *Canadian Journal of Economics*, **14**, 450–69.

Munasinghe, M. and W. Cruz (1995), or 'Economy-wide policies and the environment, lessons from experience, *World Bank, Environment Paper* 10, Washington, DC: World Bank.

Munasinghe, M. and A. Persson (1995), 'Natural resource management and economywide policies in Costa Rica: a computable general equilibrium (CGE) approach', *The World Bank Economic Review*, **9**, 259–85.

OECD (1993), *Environmental Policies and Industrial Competitiveness*, Paris: OECD.

Office of Technology Assessment (OTA) (1992), *Trade and Environment: Conflicts and Opportunities*, Washington, DC: US Congress, OTA-BP-ITE-94.

Perroni, C. and R.M. Wigle (1994), 'International trade and environmental quality: how important are the linkages?', *Canadian Journal of Economics*, **27**(3), 551–67.

Pethig, R. (1976), 'Pollution, welfare, and environmental policy in the theory of comparative advantage', *Journal of Environmental Economics and Management*, **2**, 160–69.

Piggott, J. and J. Whalley (1992), 'Economic impacts of carbon reduction schemes: some general equilibrium estimates from a simple global model', Center for Economic Studies Working Paper 17, Munich: University of Munich.

Rauscher, M. (1991), 'National environmental policies and the effects of economic integration', *European Journal of Political Economy*, **7**, 313–29.

Rauscher, M. (1995), 'Environmental policy and international capital mobility: an aggregate view', presented to the Workshop on Environmental Capital Flight, Wageningen University, October.

Rauscher, M. (1997), 'Environmental policy and international capital movements', *International Trade, Factor Movements and the Environment*, Oxford: Oxford University Press, ch. 3.

Rutherford, T. (1992), 'Welfare effects of carbon dioxide restrictions', OECD Working Paper 112, Paris: OECD.

Samuelson, P.A. (1953), 'Prices of factors and goods in general equilibrium', *Review of Economic Studies*, **21**, 1–20.

Siebert, H. (1995), *Economics of the Environment*, 2nd and 4th edn, Berlin: Springer.

Siebert, H., J. Eichberger, R. Gronych and R. Pethig (1980), *Trade and Environment: A Theoretical Inquiry*, Amsterdam: Elsevier/North-Holland.

Smith, V.K. and J.A. Espinosa (1996), 'Environmental and trade policies: some methodological lessons', *Environment and Development Economics*, **1**, 19–40.

Steininger, K. (1995), *Trade and Environment. The Regulatory Controversy and a Theoretical and Empirical Assessment of Unilateral Environmental Action*, Heidelberg: Physica.

Tobey, J.A. (1990), 'The effects of domestic environmental policies on patterns of world trade: an empirical test', *Kyklos*, **43**, 191–209.

Tobey, J.A. (1993), 'The impact of domestic environmental policy on international trade', in H. Giersch (ed.), *Economic Progress and Environmental Concerns*, Berlin: Springer, pp. 181–200.

Ugelow, J. (1982), 'A survey of recent studies on costs of pollution control and the effects on trade', in S. Rubin (ed.), *Environment and Trade*, New Jersey: Allanheld, Osmun and Co.

Welsch, H. and F. Hoster (1995), 'A general equilibrium analysis of European carbon/energy taxation', *Zeitschrift für Wirtschafts- und Sozialwissenschaften*, **115**, 275–303.

Winters, L.A. (1984), 'Separability and the specification of foreign trade functions', *Journal of International Economics*, **17**, 239–63.

Xing, Y. and C. Kolstad (1995), 'Do lax environmental regulations attract foreign investment?', Working Paper in Economics 6–96, University of California, Santa Barbara.

10

Partial Equilibrium Models of Trade and the Environment

Kerry Krutilla

10.1 INTRODUCTION

In recent years, research exploring the nexus between environmental policy evaluation and non-strategic trade theory has increasingly relied on general equilibrium analysis. Nonethe less, partial equilibrium modelling affords an efficient means of policy evaluation in an open-economy setting if the effects of policy actions on factor incomes are not of interest and outcomes do not differ substantively from those of general equilibrium analysis.[1] In the 'trade-and-environment' context, partial equilibrium models are particularly useful for studying the consequences of terms-of-trade effects, and for indicating how such factors as a country's commodity trade balance, and the type of the externality problem, affect the normative properties of environmental policy actions.

This chapter reviews partial equilibrium modelling to assess the impact of trade liberalization on the environment, and to determine the structure of optimal environmental policy in an open-economy setting. For most of the chapter, pollution is assumed to be local, but the effects of environmental regulation in the transboundary context are also briefly considered.

A number of standard assumptions underlie the analysis. First, in keeping with the orthodox trade literature, the 'rest of the world' does not respond strategically when a country initiates environmental regulation or trade-policy reform. Second, the environmental distortion in question is the only distortion in the economy, except when other distortions are explicitly addressed within the modelling frameworks.[2] Finally, the production process is one in which emissions are proportional to output.

This assumption is stronger than necessary to ensure that economic adjustments to environmental policy actions influence commodity prices and international trade flows, but it is maintained in this form for expositional convenience.[3]

The following two sections of the chapter focus on environmental policy in the case where pollution is local and economic adjustment responses in the international trade system exclusively determine the open-economy welfare ramifications. Section 10.4 extends the analysis to consider the effects of transboundary pollution, in which the direct environmental impact channel, as well as trade linkages, influence the results. Section 10.5 offers some brief concluding remarks.

10.2   TRADE LIBERALIZATION AND THE ENVIRONMENT[4]

Environmentalists have frequently expressed concerns about the environmental impact of trade liberalization (Charnovitz, 1992; Wathen, 1993; Esty, 1994). However, the relationship between trade liberalization and environmental consequences is not straightforward, even in the simplest cases. Consider a small country with one polluting commodity sector such that $e = K_q$, where e are emissions, K is a proportionality constant, and q is the sector's output. For expositional convenience, the country is assumed not to trade in the start-point equilibrium; P, q and e are the price, quantity and emissions equilibria in autarky (Figure 10.1). Assume first that the production side of the economy

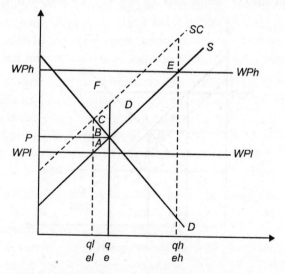

Fig. 10.1: Welfare impact on small countries from trade liberalization-production externality

272 ♦ Economics of Environment and Development

produces pollution yielding a social cost curve (*SC*) which is above
the private supply curve (*S*) (drawn linearly in Figure 10.1 for convenience).[5]
Now the country eliminates the prohibitive trade barrier, allowing unrestricted
trade at the world price (which, by assumption, the country is too
small to influence). If the country is economically positioned to be a
net exporter of the product – $P < WPh$, where *WPh* is the high world
price – production and emissions will increase, all else constant (to *qh*
and *eh*, respectively, in Figure 10.1). Without environmental policy to
redress the environmental damages, the conventionally measured welfare
benefit of the trade liberalization ($F + C + D$) is attenuated by additional
environmental damages ($D + E$) and could be negative (if $E > C + F$).
Conversely, if the country is economically positioned to be a net importer
– $P > WPl$, where *WPl* is the low world price – opening the country
to trade will reduce emissions (from *e* to *el*) as the dirty import-
competing sector contracts (from *q* to *ql*) (again see Figure 10.1). In
this case, the environmental side effect increases the welfare gain of
the trade policy (by $A + B + C$).

When the externality is generated by the product's consumption,
so that the marginal social benefits are lower than the market demand
curve (*DSB* < *D* in Figure 10.2), the welfare effects of trade liberalization,
in relation to the commodity trade balance, are just the reverse. With
the country positioned to be a net exporter, the commodity price rise
associated with the trade liberalization reduces domestic consumption
and pollution (to *ql* and *el* respectively in Figure 10.2). The environmental
impact of the lower emissions ($A + B + C$) augments the welfare

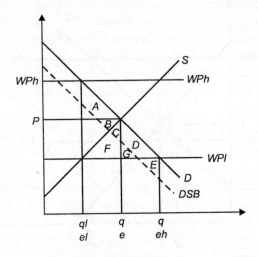

Fig. 10.2: Welfare impact on small countries from trade
liberalization-consumption externality

gains of the trade liberalization. On the other hand, when the country is positioned to be a net importer, the domestic price decline associated with the trade liberalization increases consumption and pollution (to qh and eh in Figure 10.2), reducing the welfare benefits of trade liberalization (into the negative range if $E > C + F + G$). These results are summarized in Table 10.1.

Trade liberalization is unambiguously welfare-enhancing in all cases when paired with an optimal environmental policy instrument (Anderson, 1992a). The analytics in this case do not differ in any important way from those in the case where environmental policy is implemented in a competitive closed economy. Assuming that the transactions costs of employing the appropriate environmental policy instrument are not too costly, this reality underlies the usual view of economists that environmental side effects associated with trade liberalization pose an environmental-policy problem, rather than a trade-policy problem, *per se* (see Bhagwati, 1993).

Table 10.1: Environmental Impact of Trade Liberalization
—Small-country Case

	Direction of welfare effect	
	Net exporter	*Net importer*
Production externality	−	+
Consumption externality	+	−

It is possible that the transaction costs of using the environmental policy instrument are prohibitive in some cases, for example, to ameliorate area source water pollution (Krutilla, 1997). In this circumstance, some degree of trade restraint may be theoretically justified as a second-best policy approach for the two out of four cases where trade liberalization worsens environmental damage (Anderson, 1992a).

Environmentalists have also expressed concern about the impact on small countries when larger countries liberalize trade (Charnovitz, 1992; Wathen, 1993). What happens to the environment and welfare of a small country, or a group of small countries, when their trading partners initiate environmental policy? Assume that the block of countries which regulate, at least taken together, is large enough to influence the world price. Further assume that the externality is on the production side of the same commodity market for all countries in the trade system. Given these assumptions, the environmental regulation will reduce production in the regulating countries, thereby reducing global excess supply and raising world price (Krutilla, 1991a). The transmission of the price effect will shift the production locus of the pollution, with some of the production displaced from the regulating countries to the countries which do not regulate. That is, the price effect will boost domestic output and pollution in the non-regulating countries–whether the polluting industry is export-oriented or import-competing.

The environmental regulation will also affect the terms of trade. The price increase will boost profits to exporters in the small country, which at least partially offsets the negative impact of the environmental damage. Conversely, if the country is a net importer, the increase in the cost of imports imposes a negative welfare loss which reinforces the consequences of the increased environmental damage (Krutilla, 1991a).

If the externality is on the consumption side, there is the same environmental damage impact but the terms-of-trade effect is in the opposite direction. Consumption restraints in the other countries will reduce global demand, increase global excess supply, and reduce world price (Krutilla, 1991a). Whether the non-regulating countries are net exporters or importers, the lower world price will increase consumption, worsening environmental damage. However, lower prices hurt exporters but provide a welfare gain if the country is a net importer. Thus countries which are net exporters unambiguously lose in this case, while terms-of-trade effects attenuate the environmental damages when the country is a net importer. The results of these permutations are summarized in Table 10.2.

Table 10.2: Welfare Effect on a Small Country
When Larger Countries Regulate

	Direction of welfare effect			
	Net exporter		Net importer	
	ED	TOT	ED	TOT
Production externality	–	+	–	–
Consumption externality	–	–	–	+

ED = environmental damage effect
TOT = Terms-of-trade effect

There is a certain irony to these conclusions. In two of the four cases, the terms-of-trade effects attenuate the welfare loss on the environmental side which, in any event, may be addressed with an environmental policy instrument. Thus the environmental side effect may not be a major issue in these cases. However, terms-of-trade deterioration adds to the welfare losses associated with increased environmental damage in the other two cases. Again, domestic environmental policy can be used to address the environmental side effect, but small countries have no policy instrument with which to cope with an exogenous deterioration in their terms of trade. Consequently, the terms-of-trade effects may have more of a significant impact on the welfare of small countries than additional environmental damages. This possibility does not appear to have received much notice by environmentalists and policy makers in the trade and environment area.

10.3 OPTIMAL ENVIRONMENTAL POLICY MAKING IN AN OPEN ECONOMY

In the large-country case, the terms of trade becomes a variable under the country's control. In this case, it becomes difficult to separate the welfare effects of a country's trade policies from the welfare effects of its environmental policy making. Thus we now turn to the subject of optimal environmental policy in the large-country context and its relationship to the terms of trade.

In theory, environmental impacts and terms-of-trade effects can be handled directly by using both trade and environmental policy instruments (first-best), or indirectly by using either an environmental policy instruments or tariff to address both effects (second-best). The following formulae (from Krutilla, 1991a) are useful for sorting out the permutations:

$$ts^* = \in(qs) + \alpha(qs - qd) + \alpha X_p T \qquad (10.1)$$

$$td^* + \in(qd) + \gamma(qs - qd) + \gamma X_p T \qquad (10.2)$$

where ts^* is a welfare-maximizing environmental tax on producers, holding constant the tariff T, td^* is a welfare-maximizing environmental tax on consumers, given T, qs and qd are the level of domestic production and consumption, respectively; $\in(qs)$ is the marginal environmental damage cost of production; $\in(qd)$ is the marginal damage cost of consumption; $(qs - qd)$ is excess supply; $X_p < 0$ is the slope of the rest of the world's excess demand schedule, $(X(P))$, where P = the world price; and $\alpha \geq 0$, $\gamma \leq 0$ are terms-of-trade parameters reflecting the slopes of the behavioural equations in the system (Krutilla, 1991a).[6, 7] There is a sign convention reflected in equations (10.1) and (10.2) that $T < 0$ when $qs < qd$, while $T > 0$ when $qs > qd$. We now consider some special cases.

Case 1. Small-country case: $X_p = -\infty$, $\alpha = 0$, $\gamma = 0$, $T = 0$.

This may be taken as a benchmark. With no terms-of-trade effects or tariffs, equations (10.1) and (10.2) reduce to $ts^* = \in(qs)$ and td^*

$= \in (qd)$ for the production and consumption cases respectively, that is, the standard Pigouvian taxes. This result is consistent with the earlier statement that small countries can dichotomize trade and environmental policy making by levying the optimal environmental tax, ts^* or td^*, as $T \rightarrow 0$ with trade liberalization.

Case 2. Large-country case: $X_p \neq -\infty$; $\alpha > 0$, $\gamma < 0$.

Maintaining for the moment the assumption that $T = 0$ – perhaps, for political reasons, the country has implemented a total trade liberalization– equations (10.1) and (10.2) show that the optimal (second-best) environmental taxes are:

$$ts^* = \in (qs) + \alpha(qs - qd) \qquad (10.3)$$

$$td^* = \in (qs) + \gamma(qs - qd) \qquad (10.4)$$

for the cases of production- and consumption-side externalities, respectively. These formulae differ from the classic Pigouvian tax by the presence of the terms-of-trade terms, $\alpha(qs - qd)$ and $\gamma(qs - qd)$. With $T = 0$, equations (10.3) and (10.4) indicate the formulae for a country optimally to recapture terms-of-trade losses via second-best environmental policy instruments. To do so, equation (10.3) shows that, in the production externality case, $ts^* > \in (qs)$, when $qs - qd > 0$, and vice versa. That is, environmental regulation should be higher than the Pigouvian tax level, $\in (qs)$, when the large country is a net exporter, and lower than the Pigouvian tax level when the large country is a net importer. The intuition follows from the discussion in the previous section: an environmental production restriction lowers world excess supply and raises world price, a terms-of-trade change which provides extra profits to exporters but increases the cost and volume of imports when the country is a net importer. The optimal environmental tax should be raised in the first instance to allow domestic producers to capture the additional rents, but lowered in the second instance to reduce the impact on consumers of higher-cost imports.

Since the terms-of-trade effect works in the opposite direction when consumption is taxed, the consumption tax (equation 10.2) will be higher (lower) than the Pigouvian level when excess supply is negative (positive). These results are summarized in Table 10.3.

When $T \neq 0$, equations (10.1) and (10.2) show that the tariff terms work in the opposite direction of the terms-of-trade term. Effectively, the tariff instrument handles some of the terms-of-trade effect which the terms-of-trade term would otherwise address. This relationship obtains as long as the existing tariff is lower than the optimal tariff of international trade theory (see Krutilla, 1991a). In that case, the

environmental tax can again revert to the Pigouvian tax level, and trade
and environmental policy can again be dichotomized.

Table 10.3: Welfare Effect of Environmental Regulation
in Large Countries

| | Direction of welfare effect | | | |
| | Net exporter | | Net importer | |
	ED	TOT	ED	TOT
Production externality	+	+	+	−
Consumption externality	+	−	+	+

Abbreviations as for Table 10.2.
Source: Krutilla (1991a).

Now consider the case where the environmental policy instrument
is not available, so that a second-best tariff must be used to address
both terms-of-trade and environmental effects. Letting *ts** and *td** equal
zero in equations (10.1) and (10.2) and solving for *T* we have:

$$T^* = -\frac{P}{\eta} - \frac{\in(qs)}{\alpha X_p} \qquad \text{(net exporter)} \qquad (10.5a)$$

$$-T^* = \frac{P}{\eta} + \frac{\in(qs)}{\alpha X_p} \qquad \text{(net importer)} \qquad (10.5b)$$

$$T^* = -\frac{P}{\eta} - \frac{\in(qd)}{\gamma X_p} \qquad \text{(net exporter)} \qquad (10.6a)$$

$$-T^* = \frac{P}{\eta} + \frac{\in(qd)}{\gamma X_p} \qquad \text{(net importer)} \qquad (10.6b)$$

Equation (10.5) shows the formulae for a production externality,
while equation (10.6) shows the formulae for consumption.[8]
Note than η is the elasticity of world excess demand,
$\eta = h = X_p(P/X) = X_p((P/(qs-qd)); \eta < 0$ when $qs > qd$; $\eta > 0$ when
$qs < qd$. It can be seen that the first terms in equations (10.5) and
(10.6) are the standard optimal tariffs of international trade theory (see
Krutilla, 1989). The second term adjusts the optimal tariff for environmental
damage. For the production externality, the environmental damage term
is positive when the country is a net exporter–increasing the optimal
tariff (equation 10.5a).[9] In this case, production should be restricted
more than in the standard case to reduce environmental damages; thus

raising the optimal export tariff when it is the second best means accomplishing the environmental policy objective. When the country is a net importer, environmental damages lower the optimal tariff (second-term equation 10.5b is negative). In this case the level of imports should be increased, relative to the usual situation without the environmental externalities, to reduce environmentally damaging domestic production. This requires a lower second-best optimal tariff.

The opposite relationship results when the externality is on the consumption side; the optimal tariff should be lower in the net export case in order to export more of the polluting product abroad, and higher in the net import case to restrain more domestic consumption.

Note that because subsidies have the opposite effects of taxes, all the results in this section would hold with reverse sign for positive externalities (Krutilla, 1991a). If there is a positive externality on the production side, for example, the optimal tariff would be lower when the country is a net exporter to induce more domestic production, and higher when the country is a net importer to induce more domestic production.

In reality, the GATT significantly constrains the active use of tariffs as policy instruments, and trade liberalization appears to be the widely held policy goal (Bhagwati, 1993). Hence equations (10.1) and (10.2) are probably the most empirically relevant for assessing second-best environmental policy making in the open-economy context.

10.4   TRADE AND TRANSBOUNDARY POLLUTION

When pollution spills over national boundaries, the international transmission of welfare effects occurs through an environmental impact channel as well as through price effects (see Baumol and Oates, 1988). Adopting a strictly local perspective, the stringency of a country's environmental policy should be lower, all else constant, to the degree that pollution falls out on other countries. The limiting case is when all pollution is exported and local environmental policy becomes unnecessary.

The distinction between purely local and global effects blurs when pollution produced by trading partners mutually falls out on all players in the trade system. In this case, an incentives-compatible cooperative agreement, rather than unilateral regulation, becomes the first-best policy approach. Since cooperative agreements involving all the relevant players may be difficult to achieve in practice, however, it is worth examining the factors which determine the effectiveness of less-than-comprehensively implemented environmental policy.

Consider a global pollution problem and two subsets of nations: a group which has agreed collectively to curtail emissions (which may be referred to as the 'home' countries) and a group which refuses to cooperate to reduce emissions (which may be referred to as the 'abroad' countries). Imagine that the world's supply capacity of the polluting product is exclusively located abroad, and exclusively consumed by the home countries. In this case, the home countries could reduce emissions entirely by phasing out imports of the polluting product. Using an import tariff as the policy instrument would transfer rents to the home countries during the initial phase-out period, providing compensation to home countries for the environmental regulation while penalizing abroad countries for non-compliance (Krutilla, 1991b).[10]

If the market dichotomy between home consumers and abroad producers is less than total, but the additional supply capacity in the home countries is still small enough so that home countries maintain their status as net importers, $(qs - qd < 0)$, a home consumption tax becomes the best instrument to reduce emissions. Such a consumption restraint would increase excess supply on the world market (see Sections 10.2 and 10.3), and the resulting price signal would reduce emissions production abroad, reinforcing the policy objective. The regulating countries will ultimately phase out all consumption at home, reducing production abroad to the point where it is determined exclusively by the size of the residual domestic market. If the residual market is small, the erosion of the policy through non-participation would be relatively slight—as was the case with the Montreal Protocol.[11] On the other hand, if the residual market is large, the recalcitrant nations, through their non-participation, could significantly diminish the policy's effectiveness. The size of the domestic market for coal consumption in India and China offers a good example; these countries, unilaterally or in combination, have the capacity significantly to undermine international efforts to restrain carbon emissions.

Imagine now that there is enough supply capacity of the polluting product at home so that the home countries are in fact net exporters $(qs - qd > 0)$. Interestingly, this situation diminishes the effectiveness of less-than-comprehensive policy during the implementation period (Krutilla, 1991b). A production tax or other restraint will reduce excess supply on the world market and raise world price (see Sections 10.2 and 10.3), initially providing an incentive to increase production and pollution abroad. This side effect undermines the policy objective during the implementation period. However, as the production restriction in the home market continues, the home and abroad markets will ultimately become dichotomized, and the volume of emissions production abroad will again be determined by the size of the residual market.

10.5 CONCLUSION

Partial equilibrium modelling yields diverse and interesting results with respect to trade and environmental policy outcomes. Trade liberalization will always be beneficial for small countries when paired with the appropriate environmental policy instrument. Without associated environmental policy, however, liberalizing trade will improve the environment in two out of four special cases, while worsening it in two others. Terms-of-trade effects in an open-economy setting are potentially significant, both for small countries which are passively subject to world price shocks when larger countries initiate environmental regulation, and for larger countries which can strategically manipulate the world price as a part of their environmental policy making. The former issue has not received much attention in the trade and environment literature.

Price signals transmitted through the world trade system have the potential to undermine or reinforce the objectives of less-than-comprehensively implemented global environmental policy during the implementation period. The direction of price signals is determined by whether the regulating block of countries are net importers or net exporters of the environmentally damaging product in the pre-regulation *status quo*. In either case, the size of the residual markets in the non-regulating countries will ultimately determine the degree of policy erosion through non-participation. Since fundamental economic factors shape the size of the residual markets, policy makers ultimately have limited scope unilaterally to influence the global consequence of less-than-comprehensively implemented environmental agreements.

End Notes

1. In fact, the partial analysis employed in this chapter can easily be translated into a simple two-good, two-factor general equilibrium framework (see Corden, 1980, p. 23).
2. See Anderson (1992b) and van Beers and van den Bergh (1997) for empirical studies of environmental policy levied in distorted markets.
3. It is only necessary to assume that emissions and other inputs into production or consumption are never completely substitutable, so that there is always some degree of linkage between the volume of output or consumption and the level of pollution (Krutilla, 1991a).

4. The first part of this section draws significantly on insights in Anderson (1992a).

5. I refer to this case as a 'production externality', while a 'consumption externality' will refer to demand-side-generated pollution. This usage contrasts with a typology which, independently of the source of the pollution, labels externalities in terms of whether they impact utility or production functions.

6. Specifically, $\alpha = [X_p + qd_{pd}]^1$, and $g = [X_p - qs_{ps}]^{-1}$ where $qd(pd)$ is domestic demand; $qs(ps)$ is domestic supply; ps and pd are the domestic demand and supply prices; and qd_{pd} and qs_{ps} are the first partial derivatives of demand and supply with respect to the supply and demand prices, $qd_{pd} < 0$; $qs_{ps} > 0$.

7. In the model from which (10.1) and (10.2) are derived, emissions are directly proportional to output, as has been assumed in this chapter. Hence production and consumption taxes are the relevant environmental policy instruments.

8. Given the sign convention previously noted, equations 10.5a and 10.6b are multiplied by -1 to yield positive magnitudes when countries are net importers (See Krutilla, 1991a).

9. The term is positive because a > 0 while $X_p < 0$.

10. This policy could possibly be justified under the GATT as a consumption tax which happens to strike imports only, because there is no domestic production.

11. When the agreement was ratified, roughly 85 per cent of the world's CFC (chlorofluorocarbon) production capacity was located in the signatory countries, leaving a residual 15 per cent that was unconstrained (Krutilla, 1991b).

References

Anderson, K. (1992a), 'The standard welfare economics of policies affecting trade with the environment', in K. Anderson and R. Blackhurst (eds), *The Greening of Word Trade Issues*, Ann Arbor: University of Michigan Press.

Anderson, K. (1992b), 'Effects on the environment and welfare of liberalizing world trade: the cases of coal and food', in K. Anderson and R. Blackhurst (eds), *The Greening of World Trade Issues*, Ann Arbor: University of Michigan Press.

Baumol, W.J. and W.E. Oates, (1988), *The Theory of Environmental Policy*, Cambridge, UK: Cambridge University Press.

Beers, C. van and J. van den Bergh (1997), 'An empirical multi-country analysis of the impact of environmental regulations on foreign trade flows', *Kyklos*, **50**, 29–46.

Bhagwati, J. (1993), 'Trade and the environment, false conflict?' in D. Zaelke, P. Orbusch and R.F. Housman (eds), *Trade and the Environment: Law, Economics, and Policy*, Washington, DC: Island Press.

Charnovitz, S. (1992), 'GATT and the environment: examining the issues, *International Environmental Affairs*, **4**, 213–14.

Corden, W.M. (1980), *Trade Policy and Economic Welfare*, Oxford: Oxford University Press.

Esty, D.C. (1994), *Greening the GATT: Trade, the Environment, and the Future*, Washington, DC: Institute for International Economics.

Krutilla, K. (1989), 'Tariff burdens and optimal tariffs under alternative transportation costs and market structures', *Economics Letters*, **31**, 381–6.

Krutilla, K. (1991a), 'Environmental regulation in an open economy', *Journal of Environmental Economics and Management*, **20**, 127–42.

Krutilla, K. (1991b), 'Unilateral environmental policy in the global commons', *Policy Studies Journal*, **19**, 126–39.

Krutilla, K. (1997), 'World trade, the GATT, and the environment', in L. Caldwell and R.V. Bartlett (eds), *Environmental Policy: Transnational Issues and Environmental Trends*, Greenwood, CT: Greenwood Publishing.

Ostrom, E., R. Gardner and J. Walker (eds), *Rules, Games, and Common-Pool Resources*, Ann Arbor: University of Michigan Press.

Wathen, T (1993), 'The guide to trade and the environment', in D. Zaelke, P. Orbusch and R.F. Housman (eds), *Trade and the Environment: Law, Economics, and Policy*, Washington, DC: Island Press.

11

Environmental Policy in Open Economies

*Michael Rauscher**

11.1 INTRODUCTION

In the 1970s and early 1980s, environmental economics was concerned mainly with the regulation of environmental externalities in closed economies. At that time, only very few authors looked at the interactions of environmental policies and international trade, for example, Baumol (1971), Markusen (1975), Pethig (1976), Siebert (1977, 1979), Siebert *et al.* (1980), and McGuire (1982). Recently, the interest in this area of research has been revived, for at least two reasons. On the one hand, the rapid growth of international trade and the ratification of new trade agreements (for example NAFTA) and the deepening of existing ones (such as the Maastricht Treaty) have increased the economic interdependences of countries. Industry lobbies fear that tight environmental standards undermine the competitiveness of their products in international markets. Environmentalists are worried that pollution-intensive industries might migrate to pollution havens and that a race for the bottom in environmental regulation will be started. On the other hand, the global dimension of environmental disruption has become an issue of increasing concern and this had not been considered in most of the early models. The revival of the trade-and-environment debate has produced a large number of publications. Major contributions are Merrifield (1988), Krutilla (1991), Anderson and Blackhurst (1992), Esty (1994), Copeland and Taylor (1994), Ulph (1996), and Rauscher (1994, 1995, 1997).

---

* I am indebted to Karl Steininger and Jeroen van den Bergh for helpful suggestions. The usual disclaimer applies.

This survey is organized as follows. The next section discusses how environmental policies affect the patterns of trade. Then I shall address the issue of using environmental policies to achieve trade-policy objectives. Afterwards I shall discuss trade interventions that are implemented to achieve environmental goals. The problem of regulatory capture of environmental policies in open economies will be addressed in Section 11.5. Section 11.6 will deal with international agreements on trade and the environment and the final section will raise some questions that still remain to be answered by trade and environmental economists. In this chapter, the term 'openness' will be used merely in the sense of openness to foreign trade in goods but not to foreign direct investment.

11.2   ENVIRONMENTAL POLICIES AND THE PATTERNS OF TRADE

According to the results of Heckscher–Ohlin trade theory, a country exports the goods that use relatively intensively the factors of production with which the country is well endowed. Besides the traditional factors of production (capital, labour and land), environmental resources enter the arena. Environmental resources are air and water quality, soil purity, the capacity of nature to assimilate wastes and toxic substances and so on. One can interpret emissions as that part of environmental resources which is used up during the production process and returned to the environment in the form of waste and pollutants. A country well endowed with environmental resources is expected to be an exporter of environmentally intensively produced goods. When is a country well endowed with environmental resources?

- One determinant is the degree of physical scarcity of environmental resources. Its components are, among others, assimilative capacity and population density. In a densely populated country, the endowment of environmental resources per capita is relatively small. Or, to put it the other way round: in a densely populated country the same level of deterioration of environmental resources affects a larger number of people than in a country with low population density.

- The second determinant is the willingness of the population to pay for conservation of environmental resources. The larger the environmental concern, the less abundant is the environmental resource and the more likely that the environmentally intensive good will be imported.

- Demand for final goods is the third determinant of factor abundance. A high level of domestic demand for an environmentally intensive good makes environmental resources relatively scarce and forces the country to import these goods after the move from autarchy to free trade.

Physical availability and environmental concern are translated by the political process into the environmental regulation which ultimately determines a country's relative abundance of environmental resources and its position in the international division of labour. Of course, the political process may lead to a biased representation of environmental scarcity, in particular when there are strong lobbies or if a high level of inequality leads to a large difference between the will of the median voter and that of the average individual in society. To make a distinction between the scarcity determined by physical factors and preferences on the one hand and by the political process on the other, Rauscher (1997) uses the terms 'true endowment' and 'de facto endowment'.

Additional considerations become relevant (i) if there are many goods and factors, (ii) if emissions have external effects on production, (iii) if consumption externalities are considered, and (iv) if trade in pollutants, for example toxic waste, is possible. In the case of many goods and factors, the prediction of the standard two-good, two-factor Heckscher–Ohlin model holds only on average: a country with scarce environmental resources may export some environmentally intensively produced goods, but not most of them; see Deardorff (1982). If environmental damages affect production, then a Ricardian productivity argument needs to be added. Productivity losses due to environmental disruption may differ across industries. Thus a country with stringent environmental policies tends to export goods whose production is particularly sensitive to pollution. If domestic use of a particular commodity is restricted for environmental reasons, domestic demand is low and the commodity will be exported. An example is the production of toxic herbicides in countries where their use is prohibited. Finally, as regards exports of hazardous waste, producers in countries with stringent environmental policies are willing to pay for the possibility to export pollutants. Thus tight regulation tends to induce waste exports.

How important is the impact of environmental regulation on international trade empirically? Not much evidence has been found up to now. The literature is reviewed in Rauscher (1997) and the conclusion is that there is not much evidence of a close relationship between trade and environmental regulation. The recent paper by van Beers and van den Bergh (1997) is a notable exception and it shows that more empirical research is necessary. The main reason for the insignificant results in many studies is that the costs of environmental regulation are rather modest in most sectors of the economy: about 2 per cent of total production costs. However, this may change if local and global pollution become increasingly severe and environmental standards are tightened in the future.

Should environmental standards be harmonized internationally? The general answer is no. Differences in endowments constitute the basis of mutual gains from trade and they should not be eliminated artificially. Harmonization is necessary only in cases where the impact of pollution is independent of the location of the source of pollution, that is, in the case of global pollutants such as greenhouse gases or substances that deplete the ozone layer.

The catchword of 'environmental dumping' has been used in the public discussion to describe situations in which a country uses laxer environmental policies than its trading partners. Used in this sense, environmental dumping represents something which is regarded as a non-problem from the point of view of trade economists. It appears to be more useful to define environmental dumping as an environmental policy which, for the sake of achieving trade-policy objectives, internalizes domestic social costs only incompletely; see Rauscher (1994). This is the subject of the following section.

11.3 ENVIRONMENTAL POLICIES AS TRADE POLICIES

In a first-best world, trade-policy instruments such as import tariffs and export subsidies should be used to achieve trade-related policy objectives. However, these instruments are not always available to the policy maker. Regional and international trade agreements restrict their use. Therefore, policy makers may think about alternative instruments and one candidate is environmental regulation. The most interesting questions in this context are (i) whether there will be environmental dumping and (ii) whether a race for the bottom is started when other countries retaliate. The following policy objectives may be thought of.

11.3.1 Terms-of-trade Considerations

A large country can improve its terms of trade by reducing the domestic demand for the import good and increasing its demand for the export good. The first-best instrument is the optimal tariff. If this is not available, environmental policies may be used. If production is environmentally harmful, the country exporting the 'dirty' good should use a stringent environmental policy since this increases the scarcity and the relative price of its export good. This is just the opposite of environmental dumping. The importer of the 'dirty' good should relax its standards and choose an environmental dumping strategy in order to reduce the price of its import good. In a Nash equilibrium where both countries choose their optimal responses, there is no race to the bottom: one country benefits from lax standards, but the other one prefers stringent environmental policies. Similar results are obtained if consumption is environmentally harmful. Laxer environmental standards increase the demand for and the relative price of the environmentally harmful good. This is beneficial to the exporter of this commodity and bad for its importer. Again, there is no race to the bottom since one country strives for tight environmental standards. If external effects on production are considered, anything can happen, depending on how these externalities are modelled; see Rauscher (1997).

11.3.2 Industrial Policies

Environmental policies can be used to support particular industries in an economy. There are two motives for doing this. The first one is welfare maximization. It is seen that no unambiguous policy implications concerning the stringency of environmental policy can be derived. The second motive is the protection of idiosyncratic interest groups from foreign competition. This can be achieved by indirect subsidies to domestic producers by means of lax environmental policies or by raising rivals' costs, for example through discriminative product standards. There is a tendency towards lax environmental regulation at home and too-strict requirements for foreign goods; see Rauscher (1997).

11.3.3 Leakage Effects and Transfrontier Pollution

Changes in environmental policies lead to changes in comparative advantage and specialization. If a country uses tighter standards, then environmentally intensive industries abroad become more competitive and tend to expand their outputs. This results in higher emission levels and tends to dilute the intended effects of the domestic policies. Since the net benefits from stringent environmental policies are reduced by the increase in foreign emissions, the optimal emission taxes will not internalize the social costs of pollution. In the Nash equilibrium emission taxes tend to be too low in all countries.

The terms-of-trade argument is probably of minor relevance since the impact of environmental policy on international trade is rather small empirically. However, the other arguments remain valid.

11.4 GREEN TRADE INTERVENTIONS

Should trade be restricted for environmental reasons? As long as environmental damages are purely national and transport externalities are neglected, the answer is no. It can be shown, however, that in second-best situations with non-optimal environmental policies a country may benefit from trade restrictions but the first-best solution would be to correct the environmental policies.

Transport externalities require the environmental regulation of transport activities, for example the taxation of transportation fuels. Of course this is a barrier to trade but a beneficial one since it removes indirect subsidies to transportation. There are other cases where trade is very close to the source of the environmental problem, for example the cases of toxic-waste exports and trade in endangered species. In these cases, trade interventions are close to first best from an economic-theory point of view.

Leakage problems constitute a second motive for trade interventions. If domestic environmental standards change the patterns of specialization such that domestic production is substituted by imported goods that are produced environmentally intensively abroad, then restrictions on imports are welfare-improving. If the domestic economy is an exporter of these goods, export subsidies are required to avoid an increase in foreign emissions. Matters are different if primary rather than final goods are traded. Probably the most severe leakage problems arise in international energy markets. CO_2 taxes imposed unilaterally by a large country reduce the world demand for energy and the decline in prices leads to an increase in energy use elsewhere. The appropriate accompanying measures of an environmental policy that restricts domestic energy use are subsidies for imports of energy or taxes on energy exports. These instruments raise world energy demand or reduce world energy supply and thus partially offset the adverse effects of the reduction of domestic energy demand. It should be noted that trade interventions designed to deal with leakage effects are optimal only from the point of view of a single country. They reduce leakage at the cost of distorting international trade. A globally optimal solution would be a combination of free trade and cooperative environmental policies involving higher emission taxes and, possibly, side payments.

Finally, trade restrictions may be used as sanctions to stabilize international environmental agreements. An example is the Montreal

Protocol on Substances that Deplete the Ozone Layer.

11.5 REGULATORY CAPTURE

Particularly in open economies, environmental regulation is subject to regulatory capture. Interest groups striving for protection from international competition lobby for lax environmental standards, for environmental tariffs, and for discriminatory product standards. Thus environmental protection may be turned into environmental protectionism rather easily. How can this be avoided?

One component of a solution is the use of market-oriented environmental policy instruments instead of command-and-control solutions. The command-and-control approach usually regulates polluters. Polluters then have strong incentives to negotiate favourable conditions, often on the basis of information to which the regulator has limited access, for example concerning the best available technology. Market-based instruments usually address polluting substances rather than polluters. Thus the group of regulatees is larger and more heterogeneous than in the command-and-control scenario. This raises the costs of creating a lobby. Moreover, the required information is marginal environmental damage and there are no inherent informational advantages on the part of the regulatees.

The other major component is monitoring. The protectionist content of environmental regulations should be documented. Supranational bodies such as the World Trade Organization (WTO) or the European Court of Justice can play a significant role in this process.

Finally, environmental trade restrictions should probably not be used unilaterally by a single country but, if they are, this should only be as a means of last resort. An international negotiation process links green trade interventions to mutually agreeable conditions and procedures and it is to be expected that the influence of national protectionist lobbies on the decision process is less than in the case of unilateral interventions.

11.6 INTERNATIONAL AGREEMENTS ON TRADE AND THE ENVIRONMENT

The policy debates of the recent past have led to the impression that international trade agreements and environmental agreements are in conflict. See Esty (1994) for a comprehensive inquiry of the issues and Rauscher (1997) for a shorter discussion.

It is widely argued that trade agreements are generally pro free trade and tend to restrict their signatory parties in their choice of environmental policies. An example is the General Agreement on Tariffs and Trade (GATT), which in its Article XX states that measures to protect human, animal or plant life or health can be adopted or enforced as long as they do not constitute a means of unjustifiable discrimination or a disguised trade restriction. Similar articles can be found in other agreements such as the Treaty of Rome and the North American Free Trade Agreement (NAFTA). Recent GATT panel decisions interpreted this article in a way which, environmentalists argue, gives precedence to free trade over environmental concerns. It should, however, be noted that this article does not explicitly prohibit green barriers to trade, nor does it say that the humans, animals or plants to be protected have to be located in the territory of the country adopting the measures. Thus extraterritorialism, albeit in conflict with some interpretations of Article XX, is not excluded generally. The NAFTA, being the most recent major trade agreement, has been influenced more than any other trade agreement by environmental concerns. It explicitly allows for green barriers to trade if these are implemented on the basis of an international environmental agreement. Moreover, it states that signatory parties should not use lax environmental standards to attract foreign direct investment. Nonetheless, there are still restrictions on the use of environmental policies. They may not be unjustified or discriminatory barriers to trade or a means of disguised protection.

Many international environmental agreements allow for or even require trade restrictions. Examples are the Convention on International Trade in Endangered Species (CITES), the Montreal Protocol on Substances that Deplete the Ozone Layer, and the Basel Convention on the Control of Transboundary Movements of Hazardous Wastes and their Disposal. The trade restrictions contained in these agreements are in conflict with the pro free-trade interpretations of the GATT. They are discriminating in that non-signatory parties are treated differently from signatory parties and one may argue that some of these measures are unjustified because there exist alternative policy instruments that are less distorting.

Can these conflicts be resolved? A less restrictive interpretation of the GATT rules may be a step in the right direction. The WTO expresses the desirability of sustainable development in its preamble and places environmental issues on its agenda. Nonetheless, one should keep in mind that, though far from perfect, the GATT has functioned rather well over half a century in removing barriers to trade and improving the global division of labour. These gains should not be gambled away by opening a Pandora's box of green protectionism.

As far as international transport is concerned, the WTO may be the right forum to negotiate an agreement on the internalization of transport externalities. In particular, an appropriate taxation of aviation fuels is difficult to implement on a unilateral basis, and an international agreement appears to be necessary to internalize the environmental costs of airborne transportation.

11.7 SUMMARY AND SOME OPEN QUESTIONS

What can be learnt from the literature on environmental policies in open economies? The best of all possible worlds is characterized by free trade and a tight environmental regulation which internalizes all environmental costs of production, transportation and consumption – including those costs that occur across the border. However, the real world is still far from the first best. Depending on the type of distortion prevailing, any kind of results can be derived. In second-best worlds (and in nth-best worlds, too, of course), environmental policies do not correctly internalize the social costs, and trade interventions become justifiable. From the empirical results, however, one can infer that most of the effects of foreign trade on environmental quality are probably rather small.

Although much has been achieved in the recent research on trade and the environment, there are still some areas where future research will offer valuable new insights. First, most of the models used in the literature are based on the assumption of atomistic competition or of oligopolies with a fixed number of firms. This is not particularly satisfying and models with endogenous market structure are desirable. Second, the problems of capture of environmental policies in open economies by idiosyncratic interest groups is not yet understood well enough and future research should be directed to the search for 'capture-proof' institutions. Third, there is a need to combine trade models with game-theoretic approaches to analyse international environmental agreements. The tying of issues, which is so relevant in practice, has not yet been analysed in depth by economic theorists. Last but not least, additional empirical research is necessary to assess the magnitude and relevance of the effects derived in theoretical models.

References

Anderson, K. and R. Blackhurst (eds) (1992), *The Greening of World Trade Issues*, New York: Harvester Wheatsheaf.
Baumol, W.J. (1971), *Environmental Protection, International Spillovers,*

and Trade, Stockholm: Alqvist and Wicksell.

Copeland, B.R. and M.S. Taylor (1994), 'North-South trade and the environment', *Quarterly Journal of Economics*, **109**, 755–87.

Deardorff, A.V., (1982), 'The general validity of the Heckscher–Ohlin theorem', *American Economic Review*, **72**, 683–94.

Esty, D.C. (1994), *Greening the GATT: Trade, Environment, and the Future*, Washington, DC: Institute for International Economics.

Krutilla, K. (1991), 'Environmental regulation in an open economy', *Journal of Environmental Economics and Management*, **10**, 127–42.

Markusen, J.R. (1975), 'International externalities and optimal tax structures', *Journal of International Economics*, **5**, 15–29.

McGuire, M.C. (1982). 'Regulation, factor rewards, and international trade', *Journal of Public Economics*, **17**, 335–54.

Merrifield, J.D. (1988), 'The impact of abatement strategies on transnational pollution, the terms of trade, and factor rewards: a general equilibrium approach', *Journal of Environmental Economics and Management*, **15**, 259–84.

Pethig, R. (1976), 'Pollution, welfare, and environmental policy in the theory of comparative advantage', *Journal of Environmental Economics and Management*, **2**, 160–69.

Rauscher, M. (1994), 'On ecological dumping', *Oxford Economic Papers*, **46**, 822–40.

Rauscher, M. (1995), 'Environmental legislation as a tool of trade policy', in G. Boero and Z.A. Silberston (eds), *Environmental Economics: Proceedings of a Conference held by the Confederation of European Economic Associations at Oxford, 1993 CEEA Conference*, Basingstoke: Macmillan, pp. 73–90.

Rauscher, M. (1997), *International Trade, Factor Movements, and the Environment*, Oxford: Oxford University Press.

Siebert, H. (1977), 'Environmental quality and the gains from trade', *Kyklos*, **30**, 657–73.

Siebert, H. (1979), 'Environmental policy in the two-country case', *Zeitschrift für Nationalökonomie*, **39**, 259–74.

Siebert, H., J. Eichberger, R. Gronych and R. Pethig (1980), *Trade and the Environment: A Theoretical Enquiry*, Amsterdam: North-Holland.

Ulph, A. (1996), 'Environmental policy and international trade: a survey of recent economic analysis', in H. Folmer (ed.), *International Yearbook of Environmental Economics*, Cheltenham, UK and Brookfield, US: Edward Elgar, pp. 205–42.

van Beers, C. and J.C.J.M. van den Bergh (1997), 'An empirical multi-country analysis of the impact of environmental regulations on foreign trade flows', *Kyklos*, **50**, 29–46.

12

Environmental Management in Business Firms

Pushpam Kumar

12.1 INTRODUCTION

It is of critical importance that the basis of planning and other decision-making processes take cognizance of environmental implications of various activities in the economy. The environmental impact of economic activities is felt in every sector *i.e.* agriculture, industry and services but the bulk of the GDP is derived from the manufacturing and construction sectors. Therefore, environmental environmental. Also, The management at firm level commands special place. Business activities have been one set of activities, which is ever growing, and unless the environmental objectives are incorporated and addressed at every step of the business of the firm- right from production to after sale services, the goal of sustainability would remain incomplete. Business and industries both in, the private sector as well as public sector are often portrayed simply as an exploiter of natural resources (*e.g.* forestry by paper and pulp industry) and polluter of environmental quality (*e.g.* atmospheric pollution by refineries and other chemical industries). However, as a major user of nature and its services, business and industry can play an important role in the protection of nature and environment. The private sector acts at three levels: in partnership with governments, in partnership with other stakeholders and on its own.

Partnerships between the private sector and other stakeholders often evolve from situations that are initially confrontational. An example of such an evolution is cooperation between BHP (an Australian oil company) and The Nature Conservancy (TNC) that developed in Texas after TNC purchased land for preservation that included BHP oil and gas wells. Initially TNC made demands for site restoration that BHP thought would cause considerable disruption to its operations.

In India, business houses and manufacturing firms have started realizing the importance and usefulness of their social responsibility for the long-term viability and sustenance of their survival and growth. Compliance with the existing norms of pollution and guidelines for sustainable utilization of natural resources inclusive of soil, water, forestry and biological resources has considerably increased in last few years in India. As on date, around 400 firms have complied with the ISO14001, which can be taken as an index of social responsibility of the corporate sector (CII 2003). Some of the compliances are under legal pressure, some under technical and some under the domain of economic incentive. Corporations fulfilling and complying with the societal goals, which are otherwise outside the domain of market process, are called as socially responsible organizations. This could be on voluntary basis or under pressure from different segments of society. Sometimes the threat perception from the society works more effectively than anything else. In many countries, emergence of green movement and consumer awareness has put a formidable pressure on the business sector to a respond to potentially threatening situation. Ultimately? This is referred to as corporate environmentalism. The genesis of Corporate Environmental Management (CEM) goes back to the issue of changing course of businesses in 1980s and 1990s where eco-efficiency was presented as a management approach to improve competitiveness and ecological efficiency. The concept was also mooted as the corporate response to the sustainability embodied in the work of the Brundtland Commission (WCED, 1987). The eco-efficiency model, as described by the World Business Council for Sustainable Development (WBCSD), requires companies to reduce the material intensity of goods and services; reduce the energy intensity of goods and services; reduce toxic dispersion; extend product recyclables; maximize sustainable use of renewable resources; extend product durability; and increase the service intensity of goods and services. The concept can be applied across sectors including chemical, transport, energy, and agriculture. This concept has been followed, and the principles of CEM has been applied in North America and Europe. But the experiences of developing countries has either not been documented properly or the experience has been quite varied.

12.2 DRIVERS OF CORPORATE SOCIAL RESPONSIBILITY

As is evident, if the firm complies with the standards and regulations it would increase its cost and might make the whole idea of successful business difficult. But, on the other hand, there could be a possibility

Table 12.3: CEM Drivers

Drive	Significance to Developing contexts
Personal ethics of individual entrepreneurs	In a number of instances it is the personal ethics of a CEO or another individual that drive the CSR agenda within a company. This alone cannot secure a sustainable organizational commitment to CSR since it depends on individual engagement.
Supply chain pressures from Northern trading partners	More companies are adopting voluntary codes of conduct driven by international financing requirement and head office reputation assurance. But there is yet to be a significant drive from southern companies to adoption or voluntary codes of conduct beyond a reaction to supply chain pressures from Northern trading partners.
Laws and regulations	Effectively enforced, law can be a significant driver to responsible behaviour. But although legal frameworks for environmental responsibility have been developed in much of the world, legal frameworks that require management of the social impact of business activities are comparatively undeveloped. In many developing countries, perceptions are strong that any kind of new regulation, stand or enforcement simply discourages foreign direct investment.
Public relations and reputation assurance	Public relations considerations and reputation management are among the strongest drivers for businesses engaging in CSR. On the one hand, companies view CSR as a strategic tool for promotion of reputation and brand value. On the other hand, its potential to generate spin at the expense of real change is criticized.
Shareholder activism and investor relations	There is little experience of shareholder activism in the developing world. In the North, portfolio investors such as pension funds have traditionally been largely ignorant of environmental and social issues. Investors in the North are, however, increasingly beginning to ask questions about the environmental and social practices of the companies that they invest in. Even so, 'responsible' investors are still too reliant on limited voluntary company reporting and questionnaires filled in by companies themselves.
Social license to operate	The notion that business needs to secure a 'social license to operate' from their stakeholders is widely touted as a significant driver for CSR. Increased time and expenditure in opening a new mine, demonstrable commitment to social advancement, and communication and cooperation with local stakeholders, account as requirements for operation in the developing world.
Sustaining key aspects of the business	Enclave industries such as mining, tourism, plantations, and agriculture often view certain social investment as critical to the success of their businesses. Building clinics to treat workers, spraying to prevent malaria outbreaks, providing education and treating water are some of he social development projects that businesses undertake. Companies that undertake these activities may create 'islands of development.' But history has demonstrated that in many cases these islands are fundamentally unsustainable because they rest on the continued profitability and investment of the businesses that fund them.

Table 12.3 *continued*

Co-operation in Development	There are increasing examples of co-operation, partnerships and legislation that the promote opportunities for social development such as public-private partnerships, decentralization and related policies such as Economic and Social Councils in Chile. It is becoming evident that so-called 'tri-sector partnerships' between businesses, NGOs and public institutions can promote more effective risk management and cost sharing whilst contributing to CSR.
Improving the business as a whole	Recognition that adoption of CSR practices has the potential to add value to business operating in the South in critical. The business imperative to manage social issues in society such as HIV and AIDs needs to be recognized.

(Source: Halina Ward, IIED, 2002)

where compliance increases its competitiveness and ultimately leads to profit, thus becoming a 'win-win' situation (Porter and Linde, 1995). Empirically, both types of cases (trade off and win-win) are present. The question arises as to why would a firm go for the environmental management practices. What drives them in that direction or what could drive them in the desired direction? This section attempts to explore these questions.

The factors which motivate organizations to be socially more responsive and environmental friendly can be broadly categorized into internal and external. While internal factors basically emanate from growth and profit motivation, *e.g.* pressure from supply chain, stakeholder's awareness etc., external factors would be legal and mandatory regulations of the government, or ethical and social dimensions of the firm looking for long-term sustainability practices, etc. Table 12.1 provides the list of factors which drive the firm for sound environmental management.

Both external and internal factors function as effective drivers of change for the firm to adopt CEM. It may be emphasized here that CEM primarily looks at those practices and management approaches (*e.g.* reporting, accounting and auditing) internalizing the external social costs and also complying with national pollution standards and norms along with initiatives for management of natural resources of the nation.

12.3 TOOLS OF CEM

In standard micro-economic theory, effect of production activities in terms of pollution and other effects on environment are widely studied. It is imperative to know the magnitude of the effect of production activities of the firms on the environment and previously hidden costs and benefits of a firm's activities. This has also happened due to an increase in government regulation and social expectation about business responsibilities. These responsibilities can be fulfilled to a larger extent through the methodological steps like environmental accounting, reporting, and auditing.

12.3.1 Corporate Environmental Accounting:

"The Corporate environmental accounting focuses on recording, analyzing and reporting environmentally induced financial impacts, whether current or future and ecological impacts of the firm's activities, whether current or future."(Lesourd and Schilizzi 2001)

Environmental accounting (EA hereafter) can also refer to the activities, methods and systems needed or used for recording, analyzing and reporting such information. It is a key informational input of for both efficient and effective environmental management. The central idea to any accounting framework is to measure the cost and benefit of a transaction. In commonly known financial accounting, monetary transfers are presented and the transactions, which do not translate into monetary transfer, are outside the scope of the accounting. But accounting in economics goes beyond those transactions that can only be presented in terms of monetary transfers. Environmental accounting is one such accounting, where costs and benefits are not monetary. For instance, water pollution caused by emission of chemicals from a firm into the nearby river does not have any direct monetary cost. However, if one looks carefully it will be apparent that pollution of the river is in itself an economic cost in terms of the health hazard that the pollution may create. EA accounts for these kinds of losses.

EA has come to the forefront from the perception that firms also have a social responsibility towards the environment. A profit simply in monetary terms is not enough to reflect the social responsibility of the firm. The new perception is that firms should also give an account of the effects of its activities on the environment. In other words, rise in expectations regarding the environmental responsibilities of business, whether the expectations emanate from government, consumers or society at large, has been a key factor leading to EA.

In this framework, environmental costs and benefits due to the activities of the firms are very important. The benefits can be said to be inflows and costs to be outflows. The environmental inflows and outflows with respect to the firm's physical production system is usually termed as the firm's *ecobalance*. However, the assessment of the ecological qualities of a given product necessarily rests upon the analysis of its environmental impact throughout all the stages of its life. This is called the *Life-Cycle Assessment* (LCA) of the product. Say for example, a firm produces cars. Now ask yourself, "Does production of cars lead to environmental pollution?" Simply by looking at the final product, car in this case, one cannot say whether its production is polluting. To know the same, one must know the process of production, *i.e.* each stage that the production process of the particular car passes through. Only when one comes to know the effect on environment of the various production stages, one can say whether the activity of the firm is polluting or not.

12.3.1.1 The Practice of Environmental Accounting:

As can be inferred from the above discussion, corporate environmental accounting may be thought of as a broadening of traditional financial accounting towards environmental concerns. More precisely, environmental accounting may be considered as a three-stage pursuit:

(1) Introducing accounting practices which incorporate environ-mental expenses and benefits.

(2) Measuring the environmental impacts of the firm's activities and products.

(3) Integrating the financial and ecological consequences of the firm's activities.

The first stage identifies, the balance sheet and profit and loss account of the company, plus the accounting increments relating to the environment. Thus, environmental liabilities find their way into the balance sheet. The second stage identifies, with respect to the firm's activities, environmental inflows and outflows in a manner similar to the management costing, where the aim is to be able to cost individual products and services, and hence the profitability of each entity (Chritophe (1995) in Lessourd and Schilizzi, 2001). Recently, the literature has suggested that such accounting information would be much easier to generate by adopting activity-based costing (ABC) or activity budgeting (ABB) (Brimson and Antos, 1999). The purpose of ABC is to show how much the firm impacts the environment. What impacts are most important and how these impacts are distributed. It would be important to note that information on certain chemicals can be withheld as they might provide competitors with hints on production process. Third stage of environmental accounting goes beyond the boundaries of the **firm and** identifies the environmental impacts of its products. It means that a firm would not only account for its production impacts, in terms of resource use and air and water pollution, but also for the use and disposal of its products.

The core issue here is that if the environmental performance is at stake, then not only financial but also environmental impact indicators are needed, such that accounting for the impact on the environment can be done. But producing information on environmental impacts may be very expensive and hence would need to be justified in terms of its contribution to improved management decisions and, ultimately, to the firm's bottom line. Another aspect is that environmental impacts are many. As it must be measured in physical terms data aggregation becomes a problem. Without appropriate aggregation, its value to decision makers will be limited. Nevertheless, the core issue is that information on impacts on environment must be able to be coupled with financial information linked to the firms' environmental management efforts and such efforts need to be seen by shareholders as justified.

12.3.2 Corporate Environmental Reporting

Another important aspect in the whole issue of CEM and the corporate social responsibility is the fundamental tension between the benefits and costs of corporate environmental disclosure. This issue can be approached, from an economic perspective, in terms of demand and supply of environmental reports. Hence, it is important to know the determinants of demand and supply of corporate environmental reporting.

12.3.2.1 Externalities as a source of Demand for Environmental Reporting

One of the key factors of demand for environmental reporting is an externality. A decision maker will generate an externality if by his or her decision or activity, he or she increases or reduces the benefits of a stakeholder without any compensation one way or another. If the stakeholders benefit (*e.g.* an environmental friendly technology that a firm has installed) and do not pay the decision maker for their gains, a *positive externality* is said to exist. If on the other hand the stakeholders suffer (*e.g.* pollution) from his or her decision or activity and are not compensated for their losses, a *negative externality* is said to exist. These probable negative and positive externalities work as sources of demand for environmental reporting. That is to say, from the environmental reports the positive and negative externalities can be discerned and compensation can be demanded accordingly. However when the activities of the firms affects a public good (*e.g.* the atmosphere due to gas emission), *i.e.* when it affects the community at large rather than an individual, then it is hard to determine to whom to pay the compensation. Also when the effect is inter-temporal *i.e.* the effect of activities of firms' (*e.g.* victims of nuclear radiation) take place in the future then it is tough to identify the stakeholders.

12.3.2.2 Economic Rationale for Supplying Environmental Reports and Associated Problems

The question of economic rationale of environmental reporting stems from the concern regarding cost and benefit of demand and supply of information. Most often, only the firm knows its technological expertise. Hence, much of its information about the environmental impacts of its activities remains private (*i.e. private information* to the firm). This is true not only of management with respect to employees and external stakeholders, but also between different levels of management within the company *i.e.* each level holds some private information. Therefore, the community or the government demands for this information, *i.e.* they demand for environmental report of the firms. The firms' supply depends on the cost of supply, not only in terms of disclosure of information but also the financial cost in doing so. Another important aspect from the demand side is that stakeholders should ask for the information depending upon the benefit of the same. Hence the rationale for environmental reporting should be judged from benefit and cost of the demand and supply of information.

One rule of supplying information is marginal cost of environmental accounting equal to marginal benefits (benefits can be defined as the avoided costs from environmental impacts). However in reality, because of *asymmetric information*[1], the supply of accounting information reflects not only the costs and benefit of the supplying firm but also those of the demanding party. Hence higher quality of information, characterized by higher accuracy and reliability, involves higher cost of acquiring it from the firm, in general. However the benefit derived from the high quality information will also be higher. Due to the *asymmetric information,* it is costly for the government or any other stakeholder to know whether the information provided by the firm is of high quality. Further, the issue is complicated by the problem of difference in what the demanders and suppliers of such information consider to be of optimal quality. This leads to strong disputes regarding accounting standards.

12.3.3 Corporate Environmental Auditing

Some kind of fixed standard for environment report is still largely lacking. This has resulted in the lack of quality criteria, including, as reviewed above, transparency, compatibility and reliability. At the same time, since industries are diverse, it is very hard for policy-makers to legislate in this field. The solution seems to be where accounting bodies try to reward and promote systems already implemented by companies and which seem to be better than others. A key instrument in the process of environmental report is the award schemes. In other words, an award scheme has been devised as for evaluation of environmental reports. The process of evaluation concentrates on the questions: "What constitutes a good environmental report?" "On what bases are the winners selected?" "What recommendations are made for improvements?" The Association of Certified Chartered Accountants (ACCA) proposes solution to these questions under the following criteria:

1. Company profile
2. Scope of report
3. Environmental management system
4. Relations with stakeholders
5. Communication format
6. Environmental impact details
7. Financial impacts of environmental management
8. Sustainable development and eco-efficiency
9. External certification and accreditation.

Each of these headings includes a list of sub-headings, as shown below:

1. Company Profile

   - Socio-economic context
   - Environmental Policy
   - Management commitment (both top and lower management)
   - Key environmental achievements
   - Key challenges, both environmental and otherwise

2. Scope of report

   - Number of business units included
   - Reporting principles (stakeholders, disclosure, audience)
   - Accounting principles (methodology)
   - Inclusion of all aspects of direct relevance
   - Topics included: environment, health, safety, ethics, sustainable development

3. Environmental Management System (EMS)

- Existence of an EMS
- Certification/accreditation of goals
- Integration of environmental management into everyday business
- Risk management
- Internal audit
- Environmental performance goals
- Compliance with environmental regulations

4. Relations with stakeholders

- Environmental policy regarding dialogue with stakeholders
- Organization of dialogue with employees, consumers, clients, suppliers and sub-contractors, administrations and NGOs
- Voluntary initiatives

5. Communication format

- Justification of environmental indicators chosen for the report
- Graphical presentation
- Ease of reading; user-friendliness
- Feedback system (prepaid envelopes, internet site)
- Diversity of approaches used.

6. Environmental impacts details

- Inputs by category of materials
- Output into the environment
- Waste outflows
- Packaging
- Transport
- Land contamination or disturbance and rehabilitation
- Presentation by activity sector
- Presentation by site
- Presentation by production stage (segment)
- Explanation of results with respect to past years and pre-defined targets

7. Financial impacts of environmental management

- Environmental costs
- Provisions for risks and charges
- Environmental investments
- Fines related to environmental mismanagement or non-compliance
- Financial impacts of government regulations
- Financial quantification of benefits
- Mixed eco-financial indicators (also comes under sustainability

and eco-efficiency)

- Future opportunities and risks; contingent liabilities

8. Sustainable development and eco-efficiency

  - Environmental design of products
  - Eco-efficiency indicators
  - Eco-financial indicators
  - Sustainability indicators
  - Sector-specific discussion on sustainable development
  - Opinion of employees and managers
  - Concern for social and ethical factors and impacts

9. External certification and accreditation

  - Verification by one or more external auditors
  - Transparency of methods utilized and their limitations
  - Report of critiques made to the company (and how it intends to meet the challenge)
  - Credibility (to the evaluator of the report)

Certain organizations also issue internal audit criteria. For example, Credit Suisse (a Swiss Bank) has produced a normalized format for environmental reports by banking institutions, which when slightly adapted, also fit insurance and other service companies. These are categorized into several sections as follows:

1. General information, environmental policy

  - Foreword (company's aims, challenges, stakeholders, targeted audience of report)
  - General information on bank (general policy, size, scale of activities, recent trends and so on)
  - Environmental policy of company (principles, commitment, strategies and so on)
  - Summary/Assessment (bird's eye view on key goals/achievements in recent years)

2. Environmental management system (EMS)

  - Organization and responsibilities (EMS diagram and staff, who does what, accountability)
  - Controlling (monitoring, intervention, urgencies, insurance, resources and so on)
  - Commitment of employees (means of trading, continuous improvements, assessment)
  - Environmental goals and measures/goal attainment (quantitative and qualitative)

3. Corporate ecology

- Methodology for obtaining data; systems boundaries (for example, any FCA?)
- Inflows/outflows of relevant materials and energy: corporate eco-balance (justification for degree of aggregation-competition, patented processes and so on)
- Environmental goals and measures/goals attainment (recycling, waste management and so on)
- Corporate ecology over the years (past trends, future projections)

4. Product ecology

- Aspects of lending/insuring and their environmental impacts (client discrimination)
- Aspects of capital investment and their environmental impacts
- Environmental goals and measures/goal attainment (eco-efficiency of clients and so on)

5. Communication and dialogue to stakeholders

- Environmental communication with stakeholders (which stakeholders? Ease of communication?)
- Feedback offers (internet, media, how is/ was feedback used and so on)

6. Summary

- Goals achieved during report term (compliance and internet targets)
- Milestones in environmental protection (new targets and the future)

12.4. ENVIRONMENTAL MANAGEMENT PRACTICES AND CORPORATE SOCIAL RESPONSIBILITY

The current global discourse on CSR emphasizes its cultural universality and benefits (Darley and Johnson 1993; Quazi and O'Brien, 2000). In practice, there are numerous obstacles to achieving corporate responsibility, particularly in many developing countries where the institutions, standards and appeals systems, which give some life to CSR in North America and Europe, are relatively weak. In a recently conducted study on Corporate Social Responsibility in India, it was found that public, by and large, think that Indian companies are not considered socially responsible, although a few established names famous for the philanthropic model of social development programmes are mentioned frequently (CSR WORLD Survey Report 2002 series). As far as multinational

Fig. 12.1

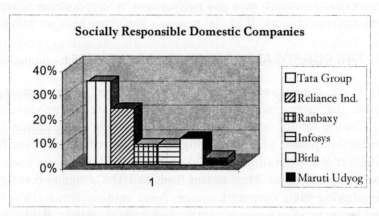

(Source: CSRWORLD 2002)

companies are concerned, they are viewed as more active in taking up CSR mainly because their efforts are more visible in the form of supply chain management and do affect a change in the scenario. In the same survey, the respondents favoured the idea of social responsibility where companies include environmental protection in their planning and strategy. Environmental responsibility featured in more than half of the replies on what CSR comprised. There was no mention of sustainable development.

The greatest name in social responsibility in India is that of the Tata Group. 31% of the respondents included Tata group in their choice of three socially responsible companies. Next to Tata Group is Reliance industry with 23%; Ranbaxy and Infosys jointly are in the third place with 8% each. The other companies mentioned are: Maruti Udyog, ICICI, ST Microelectronics, ITC, Phillips, Escorts, HCL, Hero Group and so on. This finding also confirms the global trend where the CSR has been adopted through CEM practices by some of the giant firms only.

The corporations which offer their experiences in environmental management in the form of case studies and testimonials include some of the world's most widely recognized businesses: Chevron, Dow, Dupont and Procter & Gamble (Smart, 1992 and WBCDS, 1998). Some CEOs, such as Robert Shapiro of chemicals giant Monsanto, have been become public champions of corporate environmentalism. In an interview with the Harvard Business Review, he argues that corporations can grow through sustainability; in other words, he posits that revenues may be expanded through the development of environmentally sustainable technologies and goods (Magretta, 1997). As the unprecedented number of organizations, conferences and publications by corporations illustrates, the environmental work of global business has never been so trumpeted.

12.5 THE CORPORATE RESPONSE AND ISO-14001

In India, the response to ISO-14001 has been quite encouraging. In India, the largest number of certified firms are the private Indian firms. Public Sector undertakings (PSUs), which are resource based infrastructure plants, also form a high proportion of the certified firms. Transnational Corporations (TNCs) and their affiliates have been slow to take up certification. Most Indian firms and TNCs sought certification for improving their corporate image. Joint venture firms (JVs) have, by and large, worked towards certification collaboratively. Public Sector Units (PSUs) also are pursuing certification actively. Most PSUs have got all or most of their units certified in rapid succession as a result of the clearly visible and perceived benefits of a structured response to environmental issues through EMS.For the very first company, NICCO Cables, the primary motivation for seeking certification to BS-7750 and later to ISO-14001 was the corporate image, as the CEO of the company at that time happened also to be the President of an apex industry association in the country. Subsequently, however, the company did derive both economic as well as market benefits. What helped build the favourable climate for ISO-14001 perhaps can be summarized with the help of following two major factors.

- The corporate world had already become familiar with the management system culture because of ISO-9000 success.

- The widely held perception that ISO-9000 companies do possess the capability to produce quality products, led to belief that ISO-14001 will likewise enable them to acquire the image of environmentally responsible entities.

12.6 EVIDENCES OF BEST OR BAD PRACTICE

Practices by some of the big firms around the globe eminently justify the feasibility of greater attention to CSR, while doing good business. However an unanimous conclusion seems to be difficult to arrive at. Some of the celebrated examples are:

Asea Brown Boveri (ABB), the Swedish/Swiss engineering corporation, is a sponsor of the Global Sustainable Development Facility (GSDF) and an active member of the WBCSD. It is "a world leader in developing eco-efficient technologies in a wide range of industry areas from electricity transmission to transportation, and is building a global network of joint ventures and strategic alliances to install these technologies in many developing and transition economies" (Nelson, 1996:163). ABB has also "faced sustained campaigns by environmentalists and human rights advocates against its involvement in various hydro projects, including the Three Gorges Project in China and the now indefinitely postponed Bakun Dam in Malaysia"

Aracruz Celulose, the world's largest exporter of bleached eucalyptus pulp, is often cited for its efforts to promote sustainable development through its tree planting, harvesting and pulp production processed in Brazil (Sargent and Bass, 1992). "Since its establishment, the company has earned a positive reputation both nationally and internationally for its efforts to incorporate social and environmental factors into its corporate vision" (Nelson, 1996:202). Certain investigations, however, have revealed a very different picture:"its eucalyptus trees have dried streams, destroyed the local fauna, impoverished the soil, impeded the regrowth of native plant species, and drastically reduced the area available for cultivating basic food stuffs. This is not to mention land concentration and the expulsion of the rural population." (Goncalves *et. al.* 1994, quoted in Carrere, 1999).

Dow Chemical is a US corporation which was selected to participate in the GSDF for, inter alia, "abid(ing) by the highest standards of human rights, environmental and labor, standards and norms, as defined by UN agencies" (UNDP, 1999a). According to the Transnational Resource and Action Centre (RAC)"... Dow Chemical is probably the world's largest root source of dioxin—a chlorine by-product closely associated with reproductive disorders, birth defects, increased rates of cancer, and endocrine disruption...Dow has regularly exported pesticides unregistered in the U.S for use in developing countries" (TRAC, 1999).

General Motors (GM), the world's largest automobile manufacturer, is involved in various environmental protection initiatives and partnerships. It is a Charter Partner to the Climate Wise Program, promoted by the US government to encourage energy efficiently (WBCSD, 1999:13), and in 1998 entered into a partnership with the World Resources Institute to "define a long-term vision for protecting the earth's climate and the technologies and policies for getting there" (WRI, 1998). Simultaneously, however, GM has "maintained its membership in the hard line (Global Climate Coalition)

... (which) continued to be a bastion of reaction and misinformation, and the Business Roundtable, which opposes the Kyoto Protocol..." (Frankel, 199: 11).

Mitsubishi Group has been actively cultivating an image of environmental responsibility through advertising and specific environmental projects. At the same time, it is reputed to be "a leading destroyer of tropical (and non-tropical) forests" (Greer and Bruno, 1996:182). In spring 1998, Mitsubishi was a recipient of the Corporate Watch award for companies excelling in green wash. In Mexico, the corporation has come under fire for its plans to vastly expand salt production in Baja California, which would have potentially serious implications for a local biosphere reserve, fishing communities and conflicts over land tenure and scarce natural resources such as water (Barkin, 1999)

Novartis, the Swiss life science corporation, is another participant in the GSDF and member of the WBCSD. It is often cited for its efforts in the fields of poverty alleviation and environmental protection. In 1992, for example, the company established a five-year programme to reduce the environmental and social impacts of pesticide use in the Dominican Republic (Watts and Holme, 1999:16), and the Novartis Foundation has been active in the field of corporate philanthropy (Novartis, 1998). Despite company claims to be committed to "sustainable development"' however, there are concerns that the fundamental Precautionary Principle, which the company was asked to uphold at the Earth Summit in 1992, is being ignored by virtue of the company's promotion of genetically modified crops.

Rio, Tinto, the British mining company, is often cited for its standards of environmental reporting and for promoting "continued social development" and "sustainable livelihood" in areas such as East Kalimantan, where the company's large mines approach closure (Watts and Holme, 1999:10). It has also entered into partnerships with UN agencies such as UNCTAD. Yet, according to TRAC, the corporation "has created so many environmental, human rights, and development problems that a global network of trade unions, indigenous peoples, church groups, and community activities has emerged to fight its... (Alleged) complicity in, or direct violations... in Indonesia, Papua New Guinea, Philippines, Namibia, Madagascar, the United States and Australia..." (TRAC,[2] 1999:1) (Cited in Opinion, 2002,IIED).

Business environmental requirements have begun to influence corporate behavior in favour of environmental protection. Buyers abroad are increasingly demanding proof of environmental responsibility from all those who are in the supply chain. While product related environmental requirements like eco-labels have been a part of business stipulations for quite some time, now process related certifications are also gaining currency internationally. Such certifications are related to organic and sustainable farming, floriculture, forestry, marine food etc. And such schemes are multiplying to cover all kinds of product-associated process. Just as ISO-14001 certification, all these certifications are also voluntary. The eco-consciousness of customers, however, is rapidly transforming them into a business requirement. In India, except ISO 14001, other environmental certifications are not yet very popular, although they are coming into the action gradually.

12.7 CONCLUSIONS

An important issue that emerges from the analysis is that the status and trend of environmental management practices and corporate responsibility in India and abroad is similar where only big firms are sensitive to the environmental issues. The tools of CEM like accounting, reporting and auditing can comprehensively address the concern and help the firm in becoming not only eco friendly, but sustainable for their own existence. However, most of the actions and initiatives remain largely confined to big firms in developed countries. Global agenda on the issue of CSR has somehow failed to address key development concerns of developing countries like India. This is apparent in various respects.

In the pursuit of CEM and CSR, one should not relegate some fundamental concerns which a business firms usually confronts. Firstly, the costs of corporate responsibility should not pose an excessive burden to fragile firms in developing countries that supply or compete with TNCs. Such costs need to be shared with and indeed, largely borne by TNCs and Northern consumers. Secondly, corporate responsibility should not become a protectionist tool that discriminates against Southern firms or countries. Thirdly, many developing countries have other priority concerns and need to put in place basic institutions such as the rule of law, a free press, and democratic and civil society organizations, including free trade unions. Only when such institutions exist can there be a significant progress in terms of corporate responsibility. Fourthly, there is a danger that the considerable attention to corporate responsibility issues, TNCs and their supply chains, ignores much more pressing labour, environmental and community concerns related to conditions in small and medium enterprises and the informal sector.

End Notes

1. *Asymmetric Information* refers to a situation where one party in a transaction has better information than the other. In the present case, the firm knows more about the possible impact of their technology on the environment, hence better informed about the same, than the government or the community.

2. The Transnational Resource and Action Center (TRAC), together with the Institute for Policy Studies and the Council on International and Public Affairs, prepared a report that questions the activities of various companies collaborating with UNDP in the Global Sustainable Development Facility (See TRAC, 1999)

References

Brimson, J.A. and Antos, J. (1999), Driving Value Using Activity-based Budgeting, New york: *John Wiley and Sons, Inc.*

Darley, W.K and D.M. Johnson, "Cross national comparison of consumer attitudes towards consumerism in four developing countries", *Journal of Consumer Affairs, No.27, 1993, pp. 37-54.*

Frances Carincross (1995) Green Inc.: A Guide to Business and the Environment, *London Earthscan.*

Lessourd, J. and Schilizzi, S. G. M. (2001), The Environment in Corporate Management, *Edward Elgar*, UK.

Magretta, Joan (1997), "Growth through global sustainability: AN interview with Monsanto's CEO, Robert B. Shapiro", *Harvard Business Review, January-February.*

Melody Kemp (2001) Corporate Social Responsibility in Indonesia: Quixotic Dream or Confident Expectation? *Technology, Business and Society*, United Nations Research Institute for Social Development, Geneva.

Porter, M. E. and C. Linde (1995), Toward a New Conception of the Environmental-Competitiveness Issue, *Journal of Economic Perspectives, 9(4): 97-118*

Promoting Socially Responsible Business in Developing Countries, The Potential and Limits of Voluntary Initiatives, *Report of the UNRISD Workshop 23-24* October 2000, Geneva

Quazi, A. and D. O' Brien, "An empirical test of a cross cultural model of corporate responsibility", *Journal of Business Ethics*, No.25, 2000, pp.33-51.

Smart, Bruce (1992), Beyond Compliance: A New Industry View of the Environment, *World Resources Institute: Washington, DC.*

Utting Peter (2000) Business Responsibility for Sustainable Development *United Nations Research Institute for Social Development, Geneva.*

Ward, Halina *et al.* (2002) Corporate Citizenship in opinion, IIED, London.

World Commission on Environment and Development, *Our Common Future (1987)*.

WBCSD (1998), *Annual Review* 1997: Dedicated to Making a Difference, Geneva

13

Povery and Environmental Degradation: A Review and Analysis of the Nexus

Anantha K. Duraiappah

13.1 INTRODUCTION

The poor have traditionally taken the brunt of the blame for causing society's many problems. The most recent accusation directed against them is that they cause environmental degradation. The general consensus seems to be that poverty is a major cause of environmental degradation. For example, in one of the conclusions of the Bruntland Commission report, which incidentally has been accepted as the blueprint for environmental conservation, it is explicitly stated that poverty is a major cause of environmental problems and amelioration of poverty is a necessary and central condition of any effective program to deal with environmental concerns. Along similar lines, Jalal (1993), the Asian Development Bank's chief of the environment department argues, "It is generally accepted that environmental degradation, rapid population growth and stagnant production are closely with linked the fast spread of acute poverty in many countries of Asia." The World Bank (1992) joined the consensus with its *World Development Report*, where it explicitly states, "poor families who have to meet short term needs mine the natural capital by excessive cutting of trees for firewood and failure to replace soil nutrients."

However there is a rising trend in the economic literature which disputes the conventional theory and argues that a more complex set of variables comes into play and that simple generalizations of this multidimensional problem are often erroneous and miss many important points (Leach and Mearns, 1995). There studies point out demographic, cultural and institutional factors as important variables in the poverty—environmental degradation nexus. An intricate web of these factors plus feedback loops from environmental degradation to poverty make the process of identifying causality links, if any, between these two phenomena a non-trivial exercise. These studies have been few and isolated, however, and until recently, there has been very little in-depth coordinated empirical research in the economics of environmental degradation–poverty causality relationships.

Table 13.1: Possible Relationships in the Poverty-environmental Degradation Nexus

Relationship	Description
R1	Exogenous poverty causes environmental degradation
R2	Power, wealth, and greed causes environmental degradation
R3A	Institutional failure-primary cause of environmental degradation
R3B	Market failure-primary cause of environmental degradation
R4	Environmental degradation causes poverty
R1FB	Endogenous poverty causes environmental degradation

Both poverty and environmental degradation have been increasing in many developing countries. There is pressing need to first evaluate and analyze the poverty-environmental degradation nexus and second, to prescribe policy options to mitigate or eradicate these two problems. The primary objective of the paper is to analyze critically the existing literature on the poverty-environmental degradation nexus and try to make "some order out of the chaos" inherent in this complex subject. The paper is divided into four sections. The analytical frame work developed for analyzing the poverty-environmental degradation nexus is presented in Section 2.

Section 3 is a condensed version of a more detailed literature review and analysis (Duraiappah, 1996). For this paper, we limit our analysis to the following four main natural resources which are under serious threat of degradation is many developing countries: (a) forest; (b) land; (c) water; and (d) air. We exclude biodiversity at this point because the preliminary literature search found only scattered and limited information which was too crude to contribute to the analysis of the poverty-environmental degradation link. This is an area, however, which needs particular attention in the future and the exclusion of biodiversity in this paper does not make it less important than any of the four resource sectors investigated. The paper concludes with a summary of the main findings of the literature review together with suggestions for future research.

13.2 ANALYTICAL FRAMEWORK FOR LITERATURE REVIEW

We begin by postulating a number of causality relationships which can exist between poverty and environmental degradation. To keep the analysis simple but at the same time not lose the essence of the problem, we limit our analysis to the four relationships shown in Table 13.1.

The relationships are not mutually exclusive and can be present simultaneously. Furthermore, due to the sequential nature of the relationship between poverty and environmental degradation, the following initial conditions were deemed crucial to the analysis: (a) no environmental degradation; (b) no endogenous poverty; and (c) the possibility of the existence of exogenous poverty. We define endogenous poverty as poverty caused by environmental degradation while exogenous poverty is poverty caused by factors other than environmental degradation. It can be seen that condition (b) follows from condition (a).

We begin with the popular poverty—environmental degradation relationship which states that poverty causes environmental degradation in developing countries. We call this Relationship One (R1).

R1 : Exogenous poverty
   → Environmental degradation

A counter argument to the R1 relationship is the notion that it is not poverty but a combination of greed, power and wealth that causes environmental degradation in many developing countries (Boyce, 1994). We call this Relationship Two (R2).

R2: Power, wealth, and greed
   → Environmental degradation.

A third possible relationship we could look at is that between the power/wealth/greed factor and poverty. In many developing countries, exploitation by the rich has been known to force segments of the population into poverty (Ikiara et al., 1997; Boyce, 1994). One could argue that power, wealth and greed can cause or exacerbate poverty which in turn causes environmental degradation. This could be true, but including this link in the analytical structure complicates the nexus unnecessarily because R1 should capture this phenomena during the policy prescription stage. If it is found that poverty causes environmental degradation, then the solution is to address the force causing the poverty and in that case, it would be the power/greed/wealth factor. Although, at first glance, this looks similar to the R2 relationship, the policy prescriptions are very different. In the case of the former, policy must mitigate the activities causing poverty while in the latter, policy must focus on alleviating the environment-degrading behaviour of the powerful.

The third possible relationship which we call Relationship Three A and B looks at the link between-market and institutional failures with environmental degradation respectively. In many of the previous studies on poverty and environmental degradation, the authors fail to make a distinction between market and institutional failures. In many instances, a general category called institutional failure is used to illustrate both mechanisms. This aggregation becomes unsatisfactory when policy implications and prescriptions are addressed; each failure in turn needs a unique prescription. For example, policy responses to incorrect price signals (market failure) will be quite different from policy initiatives needed to establish and enforce well defined poverty rights (institutional failure). The distinction is not always clear but it must be made if policy analysis and prescription are primary objectives (Grootaert, personal communication 1997).

R3A: Institutional failure
  $\rightarrow$ Environmental degradation.

R3B: Market failure
  $\rightarrow$ Environmental degradation

The fourth and final possible relationship which may follow either R1, R2, or R3A and R3B is the notion that environmental degradation is major factor causing poverty. This relationship is termed Relationship Four.

R4: Environmental degradation $\rightarrow$ Poverty

If R1 alone is observed then the poverty-induced environmental degradation argument can be accepted and it would be optimal from the policy makers perspective to pursue environmental protection through poverty mitigation policies. A clarification is needed at this point, however, on the type of poverty which causes the environmental degradation. From the initial conditions defined earlier, it can only be exogenous poverty which causes environmental degradation. The policies adopted should ideally be focused on the factors which are responsible for the exogenous poverty.

On the other hand, if only R2 is observed, then polices adopted under R1 assumptions can be misleading and may in fact exacerbate the degradation process as demonstrated by Binswanger (1989). But even if R2 has been rightly identified, the policy prescription may be complicated by rent-seeking activities on the part of wealthy and powerful. The first-best solution would call for the adoption of domestic policies which internalize the environmental externalities. In a majority of cases, however, especially in developing countries, vested interests could and would prevent the adoption of these solutions and second-best solutions may be the only alternative. For example, one of the many incentives for the exploitation of the natural resource base by the wealthy in developing countries is access to international markets (Chichilnisky, 1994). An option to overcome this problem could be the insistence of a standardized environmental policy, such as the Polluter Pay Principle (PPP), among trading partners. Another strategy falling under this category would be the use of international fund transfers as argued by Barbier (1990) to prevent tropical deforestation.

In the case of either R3A or R3B being responsible for environmental degradation, the solution is theoretically relatively simple; remove or correct the market or institutional failure. This may not be feasible, however. First, identifying and distinguishing between relationships R3A and R3B is challenging. Second, once the respective relationships have been identified, using policy to overcome market or institutional shortcomings is in many cases very difficult. There are many reasons for the difficulty, ranging from inertia on the part of the bureaucracy to the protection of vested interests by officials or businesses who have powerful and influential positions in the policy-making process.

Many would argue that there are critical links between R1 and R2, and institutional/market failures. We do not deny or overlook these relationships in the analytical structure used in this study. For example, if institutional or market failure is recognized as a primary reason for environmental degradation, i.e., and R3A or R3B relationship, then a address the issue, knowledge surrounding the reasons for the failure is essential before policy prescriptions can be suggested. In this process, links to R1 and R2 should be highlighted and the appropriate policy options could then be adopted.

If R4 is present, two interesting observations arise. First, R4 can only be present if it is caused by either R1, R2, R3A, R3B, or various combinations of all four. Second, the presence of R4 can set into motion an R1-type link but in this case it is endogenous poverty which causes the environmental degradation. We shall call this an R1 Feedback or R1FB link.

R1FB: Endogenous poverty
→ Environmental degradation

Let us start with the R1/R4 link. Two outcomes are possible. The first scenario would be that R1 causes R4 and the causality link ends. On the other hand, there may be a situation whereby the endogenous poverty caused by R4 sets into motion more environmental degradation by a R1FB relationship. In this instance, we get the downward spiral illustrated by During (1989). In the former, the policy strategy would be to eliminate the poverty problem at the source. In the latter, a two-pronged approach is necessary. First, and most importantly, exogenous poverty has to be addressed and stopped. Second, endogenous poverty which has been set into motion must also be addressed.

On the other hand, if R2 and R4 are present, then we are either back to a situation similar to when R2 was observed alone but with the additional presence of endogenous poverty or to a more complex situation in which the endogenous poverty caused is now itself causing environmental degradation, *i.e.*, an R1FB link. In the case where no feedback effects of endogenous poverty are present, the second-best strategies outlined in the case where only R2 is observed would be appropriate. Interestingly enough, in the more complex case whereby endogenous poverty is itself causing environmental degradation, the policy prescription may be relatively simple one; ensure that the degree or environmental degradation does not exceed the level at which endogenous poverty starts. The reasoning is as follows. By the fact that endogenous poverty can cause environmental degradation, the resource base is now under threat from two sources. The welfare of the wealthy and the powerful will decrease as the resource base they exploit is now also exploited by another group. Depending on the degree of exploitation, we may conclude that if environmental exploitation does not take place beyond the poverty "break-even"[1] level, poverty from environmental degradation (endogenous poverty) can be averted. The interesting point to ascertain is if this "break-even" point is also the "sustainable" level. Such an analysis is beyond the scope of this paper, however, and we leave this as a potential research option.

We now turn our attention to more complex situations in which R1/R2 and R3A/R3B are present simultaneously and together reinforce R4.[2] The solution to this situation is much more complex than the previous scenarios. Here, we have four contributing forces in operation: (a) the power, greed, and wealth (PGW) factor; (b) exogenous poverty (EP) factor; (c) the "institutional failure" (IF) factor; and finally (d) the "market failure" (MF) factor. Together they can be responsible for two externalities, environmental degradation and endogenous poverty. It is in fact the existence of these four factors which introduces the complex set of relationships which many of the previous studies highlight when analyzing the environmental degradation-poverty link. Moreover, because there are two externalities present, and because endogenous poverty and exogenous poverty can be distinguished, the policy prescription process is difficult and complex. For example, policies focused on the mitigation of endogenous poverty will have limited impact if the primary forces driving the environmental degradation, *i.e.*, the PGW, EP, MF and IF factors are still present. This may be one reason why many policies addressing the poverty-environmental degradation issue have failed or have had limited success.

13.3 FOREST SECTOR

Deforestation itself is not a problem and in fact may be a necessary condition for economic development. Unsustainable deforestation activities, however, result in environmental degradation. When this occurs on a large-scale, it becomes imperative to discover the factors behind the trend.

In Table 13.2, we summarize the findings from the literature review. Three activities responsible for deforestation were identified with commercial agents actively involved in all three with the poor farmer (small holdings) taking part in two of the three activities. Although no general consensus could be found with respect to the dominant activity, there was a general consensus among the studies that commercial agents were a dominant group pursuing logging and agricultural/pastoral expansion activities (Somanathan, 1991; Anderson, 1989; Repetto, 1990; Goodland, 1991; Jaganathan, 1989; Lutz and Daly, 1990; Binswanger, 1989). There was also a consensus that market and institutional failure were the main incentives driving both groups of agents to adopt unsustainable deforestation activities.

There was much less agreement among the studies on the existence of the R1 and R1FB relationships.[3] Studies disputing the R1 relationship argue that the poor do not have the resources to adopt unsustainable deforestation activities and neither do they exhibit the short time preferences which would force them to adopt the unsustainable activities (Jodha, 1990; Tiffen, 1993; Jaganathan, 1989). On the other hand, there were a number of studies which presented the opposite argument (Southgate and Pearce, 1988; Mink, 1993). The disagreement between the two groups can be reconciled to a certain extent, however, when the institutional and market failure factors were filtered out of their respective analyses. In other words, the discrepancy between the two groups narrowed

Table 13.2: Activity-relationship links for forest use

Activity	Agents	Motives	Incentives	Relationship
Logging	Commercial	Profit	Market, government policies	R2, R3A, R3B
Agricultural/pastoral	Commercial	Profit	Market, government policies	R2, R3A, R3B
	Small holdings	Subsistence	Food security	R1FB
Fuelwood	Commercial	Profit	Insecure land tenure, government policies	R2, R3A, R3B
	Small holdings	Subsistence	Basic needs	R1FB

when institutional and market structures were normalized in the analysis (Davidson *et al.,* 1992; Goodland, 1991; Lutz and Daly, 1990; Jaganathan, 1989; Southgate *et al.,* 1991; Chengappa, 1995; Browder, 1989; and Bromley and Cernea, 1989). For example, it was found that land tenure systems played a crucial role in determining the time preference factor for all groups, especially the lower income groups. The differences can be further explained by closer scrutiny of the poverty–environmental degradation link. It was revealed that endogenous poverty through the R1FB relationship was the primary factor contributing to environmental degradation activities of the marginal or poor groups (Jaganathan, 1989; Somanathan, 1991)

More detailed information, however, on the income groups responsible for the environmental degradation as well as the magnitude of their activities could shed further light on the poverty-environmental degradation nexus and help policy makers in formulating the appropriate policies to correct the situation.

The impact analysis on the other hand firmly established the existence of R4 links as shown in Table 13.3 below (Vohra, 1987; Repetto *et al.,* 1994; Bandyopadhyay, 1987; Kadekodi, 1995; Tolba *et al.,* 1992; Kumar and Hotchkiss, 1988; Duraiappah, 1994). Although the studies reviewed in this paper failed to give an indication of the magnitude of the link, it is clearly evident that welfare of agents living at the margin is being lowered by the actions of the wealthy and powerful groups (Streeten, 1994; Green, 1994; Ikiara *et al.,* 1997). Two arguments can be put forward to support urgent measures be taken to resolve the issue. First, it is well known that Pareto inefficiency implies losses to the economy in general. Second, reducing welfare of a group to the poverty level implies a further cost to the economy in general. More information and research needs to be done on linking the R1FB relationship highlighted in the activity analysis to the R4 relationships before appropriate policies can be formulated. This we believe is a major gap in the literature which needs to be addressed urgently with micro socioeconomic studies.

13.4 LAND DEGRADATION

It is estimated that 0.3 to 0.5% (5—7 million hectares) of total world arable land is lost annually due to land degradation. Dudal (1982) estimates that this figure will double by the year 2000 if present trends continue.

Unlike deforestation, the probability of one group's behaviour having

Table 13.3: Impact-relationship links for forest use

Impacts	Consequences	Groups	Relationship
Watershed protection	Rainfall disruptions, increased flooding potential	All groups affected but low-income group hardest hit	R4
Soil erosion	Productivity drop, water shortage	All groups affected but low-income groups hardest hit	R4
Destruction of safety buffer	Loss of NTFP, increased household expenditure	Low-income	R4
Productivity drop	Income drop	Low-income	R4
Fuelwood shortage	Labor productivity, increased household expenditure	Low-income	R4

Table 13.4: Activity-relationship links for land use

Activity	Agents	Motives	Incentives	Relationship
Soil exhaustion	Small holdings	Subsistence	Lack of land tenure	R1,R3A, R3B and R1FB
Soil salinization	Commercial	Profit	Water subsidy	R2, R3A, R3B
	Small holding		R3A, R3B and R1FB	
Desertification	Commercial	Profit	Lack of land tenure, government policies	R2, R3A, R3B
	Small holdings	Profit		R1FB,R3A, R3B

Table 13.5: Impact-relationship links for land use

Impacts	Consequences	Groups	Relationship
Loss of fertile top soil	Drop in agricultural productivity	All groups affected but low-income group hardest hit	R4

significant impact on another group is low for land degradation. The impact is are localized and this makes the analysis much simpler. As Table 13.4 shows, exhaustion was observed predominantly among smallholders (Southgate, 1988). The primary incentive for adopting the unsustainable activities was the lack of land tenure (institutional failure). A number of studies demonstrated quite convincingly that if secure land tenure was available, many of the poor farmers would exhibit sustainable activities and therefore soil exhaustion factors would decrease (Pagiola, 1995; Mortimore, 1989; Repetto, 1990; Mendelson, 1994; Mink, 1993; Tiffen, 1993).

In the case of soil salinization and desertification, both commercial farmers and smallholders were present within each group and were driven by the same factors. Government policies encouraging the use of water-dependent Green Revolution techniques were considered by many studies to be a primary cause for salinization (Oodit and Somonis, 1992; Repetto *et al.*, 1994). In the case of desertification, subsidies for export oriented agricultural commodities caused both groups of agents to adopt activities which eventually lead to desertification (Perkins, 1994; Unemo, 1995). If we link the results from Table 13.4 to the impact shown in Table 13.5, we can infer that the smallholders are the eventual losers from the ensuing land degradation which occurs (Jones and Wild, 1975; de Graff, 1993).

The commercial enterprises, albeit facing some losses, are able to absorb the drop in agricultural productivity in a variety of ways. On the other hand, the smallholder is restricted in his/her options to diversify the risk. In many cases, these poor farmers are pushed into the poverty group and when the initial motives of profit are replaced by subsistence demands, the R1FB relationship kicks in and further degradation of the land occurs.

An interesting observation which arises from the literature review was the link between fuelwood collection and land degradation through the fuelwood-manure link. The fuelwood shortage caused by commercial forest exploitation forces many of the poor farmers to switch to animal manure as a fuel substitute. This in turn implies less manure for soil refurbishment which inadvertently leads to soil exhaustion. This R2 link in the forest sector causes a R1FB effect in the land degradation category. This example demonstrates how complex the situation becomes when activities in one sector have feedback effects in another natural resource sector.

In conclusion, the literature analysis highlighted the existence of R2, R3A, and R3B which together produced R4 and subsequently

R1FB. Similar to the forest sector, the magnitude of the relative effects were not available. But the common theme of institutional and market failure which was evident in the forest sector was also predominant in the land sector.

13.5 WATER

The literature review identified two major issues within the water sector which play an important role in the poverty–environmental degradation nexus.

– Water shortage
– Water pollution or contamination

In its 1992 *World Development Report*, the World Bank estimates at the global level, 22 countries were facing severe water shortages while a further 18 are in the danger of facing shortages if fluctuations to the present rainfall patterns occurs. It is estimated that approximately two billion people live in areas with chronic water shortages and the numbers are expected to increase with increasing demand for water caused by growing populations and economic activity (UNFPA, 1991; Davidson *et al.,* 1992).

Although water shortage is a major threat, water contamination and pollution pose a much more immediate serious problem. Access to safe drinking water is still considered a luxury for many in the developing countries (Mink, 1993). In the past, human waste was deposited naturally in natural systems but with increasing populations, the load of human waste has far exceeded the natural systems absorption and cleansing rate. Therefore, without modern sanitation systems to help the natural systems, these systems, including water, degrade. Water contamination also comes in the form of industrial and agricultural pollutants. The cheap and easy practice of dumping industrial and agricultural effluent in lieu of expensive cleaning systems has made natural water systems a target for pollution.

As Table 13.6 shows, both commercial and small-holdings are active participants in the degradation of water resources. The motives are similar to the case of water shortage but differ significantly for water pollution. In the case of water shortages, the commercial interests were driven primarily by the PGW factor (R2) which was supported in many instances by different forms of market and institutional failure (R3A and R3B) (Oodit and Somonis, 1992; UNEP, 1995; Shah, 1993). A common theme that most of the studies point to is the absence or misuse of property rights pertaining to the use of water. Jodha (1990) as well as Singh and Balasubramanian (1977) show how in the past, village communities had very stringent rules on water use and they observed that water shortage was never the serious recurring problem that it is nowadays. With the establishment of individual property rights[4] and the breakdown of traditional institutional structures, the rights to water have quite often meant benefits to high-income groups who either had the resources to acquire the water property rights or take advantage of the access to government subsidized water supplies (Chaturvedi, 1976). In this manner, we can observe a clear R2 relationship.

In the case of the smallholders, it was the presence of water subsidies which provided the incentives to overuse the water supply (R3B). If we link these results to those shown in Table 13.7, however, we can immediately infer that the smallholdings will be more adversely affected than the commercial enterprises by the degradation. The vulnerability of the low-income groups to changes in water endowments as well as the lack of substitutability options on the part of the poor are the main reasons for the RIFB and R4 relationships being present.

In the case of water pollution, commercial agents are driven primarily by profit motives (Davidson *et al.*, 1992). On the other hand, the low-income groups pollute because of a lack of provision of proper sanitation and drinking water facilities by governmental agencies (R3A) (Leitmann, 1994). The presence of R1FB in this cases to be expected as the water degradation leaves the low-income groups no other option but to degrade further the existing water supply. This in turn causes the impacts shown in Table 13.6, which then set into motion the R4 relationship and the spiral continues (Mink, 1993; Dasgupta *et al.*, 1994; Kadekodi, 1995; Bandyopadhyay, 1987).

13.6 AIR (INDOOR AND OUTDOOR)

The World Bank estimates 1.3 billion people, most of them in developing countries, live in towns or cities which do not meet minimum WHO standards for Suspended Particulate Matter (SPM). This statistic only covers outdoor air pollution. If the coverage is extended to include the 400 to 700 million (mostly rural women and children) people exposed to unsafe levels of indoor pollution approximately two-fifths of the world's population, most of them located in developing countries, do not enjoy the basic right to clean air (Oodit and Somonis, 1992)

We can infer from Table 13.8 that a large portion of air-polluting activities are carried out by industry and higher income groups. The main motivation for these activities were profits and affluence which

Table 13.6: Activity-relationship links for water use

Activity	Agents	Motives	Incentives	Relationship
Water shortage	Commercial	Profit	Water subsidies and economies of scale	R2, R3B
	Small holdings	Survival	Lack of access	R1FB
	Small holdings	Profit	Water subsidies	R3B
Water pollution	Commercial	Profit	No pollution taxes	R2
	Small holdings	Survival	Lack of governmental support	R3A and R1FB

Table 13.7: Impact-relationship links for water use

Impacts	Consequences	Groups	Relationship
Health	Mortality increases, productivity drop	Low-income	R4
Food	Drop in protein source, productivity drop	Low-income	R4
Drought	Agricultural productivity drop	Low-income	R4

in turn were supported by a lack of policy instruments. The reverse is true in the case of indoor pollution. Low-income groups driven by the lack of access to fuel substitutes are forced to rely on highly polluting biomass fuels for heating and cooking (Tolba *et al.*, 1992; Mink, 1993). The reliance on these biomass fuels forces the low income groups to adopt unsustainable deforestation activities as illustrated by the presence of R1 and Table 13.8.

The impacts of air pollution, both indoor and outdoor, vary by income group (Leitmann, 1994). In the case of outdoor air pollution, valuation studies using hedonic methods have shown that the high-income groups can to a large extent shield themselves from the adverse impacts of air pollution (Dixon *et al.*, 1995). The low-income groups, however, are not so fortunate. In many instances, factories are normally situated in or close to low-income neighbourhoods. The resulting health consequences arising from outdoor air pollution are more prominent in the low income groups. The rise in respiratory diseases among the low income groups implies a drop in productivity which in turn forces many to lose their jobs and source of income. The ensuing drop in income, forces these groups to experience economic and social hardship which over time results in poverty. This a classic example of how environmental degradation causes poverty; the R4 relationship shown in Table 13.9.

13.7 OVERVIEW OF RESULTS

To summarize, the following factors were found to be prominent in the poverty-environmental degradation nexus.

(a) In a majority of the studies discussed above, we found that activities by the rich and powerful were the primary contributing factors forcing groups living at the margins into poverty. In other words, a combination of R2 and R4 was predominant. It is important to stress here the difference between a direct and indirect link between power and wealth with environmental degradation. R2 describes the direct link. An indirect link describes the relationship between power and wealth with environmental degradation via institutional and/or market failures. An understanding of this differentiation is critical to appropriate policy strategies.

(b) Closely related to the indirect link in (a), institutional and market failures also play a prominent role in environmental degradation and subsequently poverty enhancement: a combination of R3A/

Table 13.8: Activity-relationship links for air use

Activity	Agents	Motives	Incentives	Relationship
Outdoor pollution	Industry	Profit	No pollution taxes	R2, R3A, R3B
	Affluent groups	Affluence	No pollution taxes	R2,R3A,R3B
Indoor pollution	Low-income groups	Subsistence	Survival	R1 and R1FB

Table 13.9: Impact-relationship links for air use

Impacts	Consequences	Groups	Relationship
Drop in indoor air quality	Rise in respiratory diseases	Low-income groups, especially women and children	R4
Drop in outdoor air quality	Rise in respiratory diseases	All groups affected but low-income groups hardest hit	R4

R3B with R4. The activities of both the marginal and rich groups are influenced by these failure. In the case of the former, a combination of market and institutional failures together with lack of information were the primary reasons for adopting unsustainable practices. In the latter case, it was purely a case of exploitation—exploitation of the failures to reap maximum benefits in the shortest time horizon.

(c) The third factor prominent in the analysis is the presence of R1FB relationships. Ninety per cent of the studies show marginal groups adopting environmental-degrading activities. Of this 90%, 10% freely chose these activities. The remaining 90% had no choice but to adopt unsustainable activities. The collapse or increased vulnerability of the income stream, caused in the first instance by the activities of the powerful and wealthy, left the marginal group with few options other than to adopt resource mining activities.

(d) None of the studies reported on the losses wealthy groups accruing from the environmental degradation faced by the powerful and caused by their own activities, as well as the marginal groups. But a rapid deterioration of natural resources can only imply a worsening situation for this group in the long run. The increasing intensity of this factor coupled with the R1FB factor is becoming evident as witnessed by the increasing confrontations, in many cases violent, between the rich and the poor.

13.8 CONCLUSION

Does the literature analysis provide enough evidence to refute the hypothesis that poverty is a major cause of environmental degradation? The answer is a qualified yes, because it demonstrates without a doubt that the poor do not initially or indirectly degrade the environment. The response is qualified, because it is contingent upon the activities of other groups not degrading the environment, and an absence of market or institutional failures.

Do the powerful and wealthy degrade the environment? Again, the answer must be a qualified yes. They only degrade the environment if there are institutional or market failures. The mere fact that they can influence the market to their advantage infers some sort of institutional failure. The second condition under which this group will exhibit environment-degrading behaviour is when the marginal group begins to degrade the common environment.

From a policy perspective, two fundamental conditions must be satisfied at all times. First, institutional and market failures must be corrected. If this is not possible, then policies must be made which take into account of these imperfections. Second, groups which adopt unsustainable activities must be encouraged or given the incentives to stop. A strategy of compensation, rewards, taxes, and information provision may be needed to provide the right motivations. It is an area of research which has had little empirical work done to date and offers the potential for substantial work in the future.

End Notes

1. We define the poverty break-even point as the point at which an extra unit of environmental extraction by one agent will cause an agent who is presently just above the poverty line to fall below the poverty line.
2. We do not discuss the situation in which R3A and R3B and R4 are present as the solution to this scenario is identical to when R3B are observed alone; only in this case, the pressure to correct the institutional or market failure is higher due to the presence of two externalities—environmental degradation and endogenous poverty.
3. R1FB is endogenous poverty causing environmental degradation which was in turn caused by either R1, R2, R3A, R3B acting alone or together.
4. The establishment of individual property rights itself does not imply water shortages but the manner in which the rights were initially distributed as well as the inability or reluctance of the political and judiciary system to protect the property rights caused water shortages for the low-income groups.

References

Bandyopadhyay, J. (1987) Political ecology of drought and water scarcity: Need for an ecological water resource policy. *Economic and Political Weekly* (December 12).

Barbier, E.B. (1990) The farm level economics of soil conservation: The uplands of Java. *Land Economics* **66**(2).

Binswanger, H. (1989) Brazilian policies that encourage deforestation in the Amazon. Environment Department Working Paper, World Bank, Washington DC.

Boyee, J.K. (1994) Inequality as a cause of environmental degradation. *Ecological Economics* **11**(3).

Bromley, D. and Cernea, M. (1989) The management of common property natural resources: Some conceptual and operational fallacies. World Bank Discussion Paper No. 57, World Bank, Washington DC.

Browder, J.O. (1989) Development alternatives for tropical rain forests. In *Environment and the Poor: Development Strategies for a Common Agenda*, ed. H.J. Leonard. Transaction Books, New Brunswick NJ.

Chichilnisky, G. (1994) North-South trade and the global environment. *American Economic Review* **84**(4).

Chaturvedi, M.C. (1976) *Second India Studies: Water*. Macmillan, New Delhi, India.

Chengappa, R. (1995) Paradise. *India Today* (August 15).

Dasgupta, P., Folke, C. and Maler, K.G. (1994) The environmental resource base and human welfare. Beijer Reprint Series No. 35, Beijer Institute, Stockholm, Sweden.

Davidson, J. *et al.,* (1992) *No Time to Waste: Poverty and the Global Environment*. Oxfam, Oxford.

de Graff, J. (1993) *Soil Conservation and Sustainable Land Use: An Economic Approach*. Royal Tropical Institute, Amsterdam, The Netherlands.

Dixon, J. *et al.,* (1995) *Economic Analysis of Environmental Impacts*. Earthscan Publications, London.

Dudal, R. (1982) Land degradation in a world perspective. *Journal of soil and Water Conservation* **37**(5),245–249.

Duraiappah, A.K. (1994) A state of the art review on the socio-economics of the bamboo and rattan sector in Southeast Asia. INBAR Working Paper No.1, International Development Center (IDRG)-South Asia Office, New Delhi, India.

Duraiappah, A.K. (1996) Poverty and environmental degradation: A literature review and analysis. CREED Working Paper Series No. 8, Free University, Amsterdam.

Durning, A.B. (1989) Poverty and the environment: Reversing the downward spiral. World Watch Paper 92. World Watch, New York, November.

Goodland, R. (1991) Tropical deforestation solutions, ethics and religions. World Bank Environment Working Paper, World Bank, Washington DC.

Green, C. (1994) Poverty, population and environment: Does synergism work for women. Institute for Development Studies, Brighton, UK.

Ikiara, G. *et al.* (1997) Poverty and environmental degradation in Narok, Kenya: A background paper. Internal Working Paper, CREED, Institute for Environmental Studies, Free University, Amsterdam.

Jaganathan, N.V. (1989) Poverty, public policies and the environment. The World Bank Environment Working Paper No. 24, World Bank, Washington DC.

Jalal, K.F. (1993) Sustainable development, environment and poverty nexus. Occasional papers No. 7, Asian Development Bank, Manila.

Jodha, N.S. (1990) Rural common property resources contributions and crisis. *Economic and Political Weekly* (June 30).

Jones, M.J. and Wild, A. (1975) Soils of the West African savanna. The maintenance and importance of their fertility. Harpenden, Commonwealth Agricultural Bureau.

Kadekodi, G.K., ed. (1995) Operationalizing sustainable development, ecologyeconomy interactions at regional level. IVM Internal Publication, Institute for Environmental Studies, Amsterdam, The Netherlands.

Kumar, S.K. and Hotchkiss, D. (1988) Consequences of deforestation for women's time allocation, agricultural production and nutrition in hill areas of Nepal. IFPRI Research Report No. 69, Washington DC, IFPRI, October.

Leach, M. and Mearns, R. (1995) Poverty and environment in developing countries. An overview study. Institute for Development Studies, University of Sussex, Brighton, UK.

Leitmann, J. (1994) Rapid urban environmental assessment: Lessons from cities in the developing world. The World Bank, Washington DC.

Lutz, E. and Daly, H. (1990) Incentives, regulations and sustainable land use in Cost Rica. World Bank Environment Working Paper No. 34, World Bank, Washington DC.

Mendelson, R. (1994) Property rights and tropical deforestation. Oxford Economic Papers 46, 750–756.

Mink, S.D. (1993) Poverty, population, and the environment. World Bank Discussion Paper 189, World Bank, Washington DC.

Mortimore M. (1989) The causes, nature and rate of soil degradation in the northernmost states of Nigeria and an assessment of the role of fertilizer in counteracting the process of degradation. World Bank Environment Working Paper No. 17, World Bank, Washington DC.

Oodit, D. and Somonis, U.E. (1992) Poverty and sustainable development. In *Sustainability and Environmental Policy,* ed. F. Ditetz, U.E. Somonis and J. van der Straaten. Sigma, Berlin.

Pagiola, S. (1995) Price policy and returns to soil conservation in Kitui and Machakos, Kenya. *Environmental and Natural Resource Economics.*

Perkins, J.S. (1994) Rangeland degradation, social injustice and borehole dependent cattle keeping in the Kalahari, Botswana. In *Proceedings if the 4th International Conference on Desert Development: Sustainable Development of Our Common Future,* ed. M.A. Garduno, M.A.P. Moncayo and R.Z. Zarate, pp.237–244. Mexico City.

Repetto, R. (1990) Deforestation in the tropics. *Scientific American* **262**(4), 36–45.

Repetto, R. *et al.* (1994) The second India study revisited: Population, poverty and environmental stress over two decades. World Resources Institute, New York.

Shah, T. (1993) *Groundwater Markets and Irrigation Development—Political Economy and Practical Policy.* Oxford University Press, Bombay, India.

Singh, I. and Balasubramanian V. (1977) Effect of continuous application of chemical fertilizers on the organic matter levels of soils at Samara, Nigeria. 13th Annual Conference of the Agricultural Society of Nigeria, Zaria, Samara Conference, Institute for Agricultural Research, Zaria, Nigeria.

Somanathan, E. (1991) Deforestation, property rights and incentives in central Himalayas. *Economic and Political Weekly* (January 26).

Southgate, D. (1988) The economics of land degradation in the Third World.

World Bank Environment Department Working Paper No. 2, World Bank, Washington DC.

Southgate, D. and Pearce, D. (1988) Agricultural colonization and environmental degradation in frontier developing economies. World Bank Environment Department Working Paper No. 9, World Bank, Washington DC.

Southgate, D., Sander, J. and Ehui, S. (1991) Resource degradation in Africa and Latin America: Population pressure, policies and property arrangements. In *Arresting Renewable Resource Degradation in the Third World,* ed. D. Chapmann. World Bank Environment Working Paper No. 44, World Bank, Washington DC.

Steeten, P. (1994) Human development: Means and ends. *AEA Papers and Proceedings* **84**(2).

Tiffen, M. (1993) Productivity and environmental conservation under rapid population growth; A case study of Machakos district. *Journal of International Development* **5**(2).

Tolba, M.K. *et al.* (1992) *The World Environment, 1972–1992: Two Decades of Challenge.* Chapman and Hall, London.

Unemo, L. (1995) Environmental impact of governmental policies and external shocks in Botswana: a computable general equilibrium approach. In *Biodiversity Conservation,* ed. C.A. Perrings. Kluwer Academic Publishers, New York.

UNEP (1995) Poverty and the environment. Reconciling short term needs with long term sustainability goals. UNEP, New York.

UNFPA (1991) *Population, Resources, and the Environment: The Critical Challenges.* United Nations Population Fund, New York.

Vohra, B.B. (1987) Water resources: Land management holds the key. *The Economic Times* (New Delhi), September.

World Bank (1992) *World Development Report.* Oxford University Press, Oxford.

Index